Status of Recent Geoscience Graduates 2015

Carolyn Wilson
American Geosciences Institute
Alexandria, VA 22302

About AGI's Geoscience Student Exit Survey

The American Geosciences Institute (AGI) launched the Geoscience Student Exit Survey to assist geoscience departments in assessing the educational experiences of graduating students, as well as to examine ongoing evidence of knowledge gaps of new graduates entering the workforce. With this survey we hope to identify student decision points for entering and persisting within the geosciences, identify the geoscience research fields and co-curricular activities of interest to the students, identify the preferred jobs and industries of graduating students including those not consider part of the traditional geoscience workforce, and establish a benchmark for a detailed study of the career pathways of early career geoscientists. Likewise, the Geoscience Student Exit Survey is the gateway for an ongoing longitudinal survey of geoscience professionals that will ramp up in the spring of 2015.

The report examines the responses to AGI's Geoscience Student Exit Survey by graduates from the 2014–2015 academic year.

This survey has four major sections: student demographics, educational background, postsecondary education experiences, and post-graduation plans, with specific questions that cover areas such as community college experiences, quantitative skills, field and research experiences, internships, and details about their immediate plans for graduate school or in a new job. The survey was piloted twice in spring 2011 and spring 2012. For spring 2013 graduation, we opened the survey up to all geoscience departments in the United States. In 2014, the survey was available for graduates at the end of each semester — fall, spring, and summer, For the 2014–2015 academic year, AGI asked their federation of member societies to send the survey out to their student membership, which, in addition to the numerous department contacts already distributing the survey, reached a larger pool of recent geoscience graduates. This year AGI was also able to engage colleagues in Canada and the United Kingdom to help translate and distribute the survey for their graduates.

AGI will continue to try and expand the number of countries that distribute the survey in the future, starting with South Africa and Australia. AGI recognizes the importance of having a more global perspective of the geoscience workforce, and this survey will provide an essential piece of that understanding with information about the preparation of the future workforce.

To encourage participation, departments, member societies, and international organizations that help distribute AGI's Geoscience Student Exit Survey will receive the data in aggregate for their constituency, as long as they have a sufficient number of participating students to ensure individual response privacy.

If you would like more information or would like your department, member society, or country to participate in AGI's Geoscience Student Exit Survey, please contact Carolyn Wilson.

Status of Recent Geoscience Graduates 2015
Edited by Carolyn Wilson

ISBN: 0-913312-46-0
ISBN-13: 978-0-913312-46-9

Graphs by Carolyn Wilson, AGI Workforce Program
Design by Brenna Tobler, AGI Graphic Designer

For more information on the American Geosciences Institute and its publications check us out at
www.americangeosciences.org/pubs.

AGI
american
geosciences
institute
connecting earth, science, and people

Carolyn Wilson, Geoscience Workforce Data Analyst
American Geosciences Institute
4220 King Street, Alexandria, VA 22302
www.americangeosciences.org
cwilson@americangeosciences.org
(703) 379-2480, ext. 632

Front cover photo © L.J. Lourens: 2nd year students of the Geosciences Faculty of the Utrecht University are studying the Campanian carbonate flysch successions at Sakoneta (Basque Coast Geopark, Spain) – June 2015, back cover photo © Samantha Berkseth. All photos in this report were submitted to the 2015 Life in the Field contest, which requested images representing meaningful geoscience work through internships, research, employment, or field experiences.

GEOSCIENCE STUDENT EXIT SURVEY

Executive Summary

The American Geosciences Institute's (AGI) Status of Recent Geoscience Graduates 2015 provides an overview of the demographics, activities, and experiences of geoscience students that received their bachelor's, master's, or doctoral degree during the 2014–2015 academic year. This research draws attention to student preparation in the geosciences and their education and career path decisions, as well as examines many of the questions raised about student transitions into the workforce.

The Status of Recent Geoscience Graduates report was first released in 2013 presenting data from spring 2013 graduates. For the 2014 report, the number of participants in the AGI's Geoscience Student Exit Survey increased by 60% compared to 2013 creating a sample size that better represents the community of geoscience graduates. The 2015 report have participation rates at the approximately the same level as 2014, and many of the trends seen in 2013 and 2014 are echoed in the 2015 report.

This report presents the results for the end user's consideration. As in previous years, the quantitative skills and knowledge of the graduates have raised concerns about their preparedness for the workforce. In 2015, there was another drop in the percentage of graduates at all degree levels taking Statistics, even though a solid understanding of statistics is necessary for conducting research and reading many articles in peer-reviewed journals. Also, the majority of bachelor's and master's graduates complete Calculus II, but there is a strong drop-off in the percentages of these graduates that take higher level quantitative courses, which was also seen in previous years. Discussions with representatives from various industries have indicated that a lack in understanding of high level math can affect sustained employment or advancement depending on the job expectations.

In 2015, as in previous years, a large percentage of graduates at all degree levels indicated they did not participate in an internship. However, this year the graduates were asked how many internships applications they submitted. Their responses indicated that the graduates understand the importance of internships to their professional development, but there may not be enough opportunities available to meet the demand for these opportunities.

This year also saw some changes in the future plans of the recent graduates. Fewer graduates indicated a desire to go on to earn a graduate degree in the fall. Many geoscience departments have indicated that they are at capacity for graduate students, and the decrease in students planning to begin work on another degree may be in response to the increased competition for the slots that are available. In regards to the recent graduates entering the geoscience workforce, this was the first year since printing this report that the environmental services industry hired a higher percentage of bachelor's graduates than the oil and gas industry, indicating this industry as a strong viable option for recent bachelor's graduates in particular.

AGI is excited about the prospect of continuing this research in the coming years here in the U.S. and internationally. Soon, these reports will look at trends seen in various countries and identify the workforce preparation issues that face the U.S. versus those issues that are of global concern.

Acknowledgements

I would like to recognize a few organization and individuals for their support for this project. Thanks to ConocoPhillips for their financial support for the project this year. Thanks also to the American Geophysical Union, the American Institute of Professional Geologists, the Association for the Sciences of Limnology and Oceanography, the Geological Society of America, and the Society of Exploration Geophysicists for distributing the survey to their student membership. I also want to thank the AGI Workforce 2015 Fall Intern, Jordan Ellington, for her hard work cleaning and organizing the data from this survey, as well as her help coding the qualitative responses. Finally, I would especially like to thank the department contacts from each participating department for distributing the survey to their graduating students.

Contents

An Overview of the Demographics of the Participants

This year, AGI's Geoscience Student Exit Survey was made available to geoscience graduates at all traditional graduation periods (winter, spring, and summer) during the 2014–2015 academic year, to be collectively referenced as "2015". Approximately two months before the end of each semester, an email was sent to all the heads and chairs of geoscience departments across the country asking for their participation in this study. As incentive to participate, AGI gives the departments the data in aggregate for their graduates for their internal assessment purposes. Distribution instructions and the survey link were sent to the identified representatives for each department that agreed to send the survey to their graduates. Departments continue to have the option to customize the survey appropriately for their graduates.

For 2015, AGI asked the American Geophysical Union (AGU), the American Institute of Professional Geologists (AIPG), the Association for the Sciences of Limnology and Oceanography (ASLO), the Geological Society of America (GSA), and the Society of Exploration Geophysicists (SEG) to distribute the survey link to their student membership, which increased the participation dramatically. These societies helped to recruit approximately 38 percent of the recent graduates that participated in the survey this year.

The survey was available to the winter and summer graduates for two months, and the spring graduates had three months to complete the survey. At the close of the survey, 692 graduating students from 210 geoscience schools or departments provided responses — 495 bachelor's graduates, 127 master's graduates, and 70 doctoral graduates. All but two states, Arkansas and Delaware, are represented within this sample of geoscience graduates. This is a slight increase in participation from last year, and using AGI's graduation data from 2014, this sample size was determined as sufficient to statistically represent the total population of geoscience graduates.

The first section of the survey covered student demographics to establish an understanding of the students that graduate in the geosciences. The data remain consistent with the data collected in 2014. However, there is a shift in the gender dynamics again. In 2015, the percentage of female master's graduates over took the percentage of male master's graduates by 10 percent, and the percentage of male doctoral graduates exceeded the percentage of female doctoral graduates by 6 percent. In the previous year, there were only slightly more men than women completing a master's degree, and the women doctoral graduates exceeded the men by 11 percent. As in previous years, students indicating their citizenship as U.S. Citizen or Permanent Resident were asked to indicate their race and ethnicity. The percentage of underrepresented minorities includes African Americans, Hispanic/Latinos, Native Americans/Alaskans, and Native Hawaiians/Pacific Islanders. However, it is important to note that is the percentage of underrepresented minorities is dominated by the Hispanic/Latino population of geoscience graduates. There was a slight increase in the percentage of graduates unwilling to share their citizenship and race and ethnicity in 2015. The age distribution of graduates in 2015 is fairly similar to the age distribution of 2014 graduates.

For the 2015 survey, recent graduates were asked to report the highest education level of their parents or guardians. Concerns have been raised that geoscience programs tend to attract students from middle and upper class families, possibly due to familiarity with the subject area among family or the high cost of the activities associated with the degree. In 2015, 64 percent of bachelor's graduates, 80 percent of master's graduates, and 72 percent of doctoral graduates have at least one parent with a postsecondary degree. This question also indicated that 18 percent of bachelor's graduates, 5 percent of master's graduates, and 10 percent of doctoral graduates were first-generation college students .

Distribution of participating graduating students and departments*

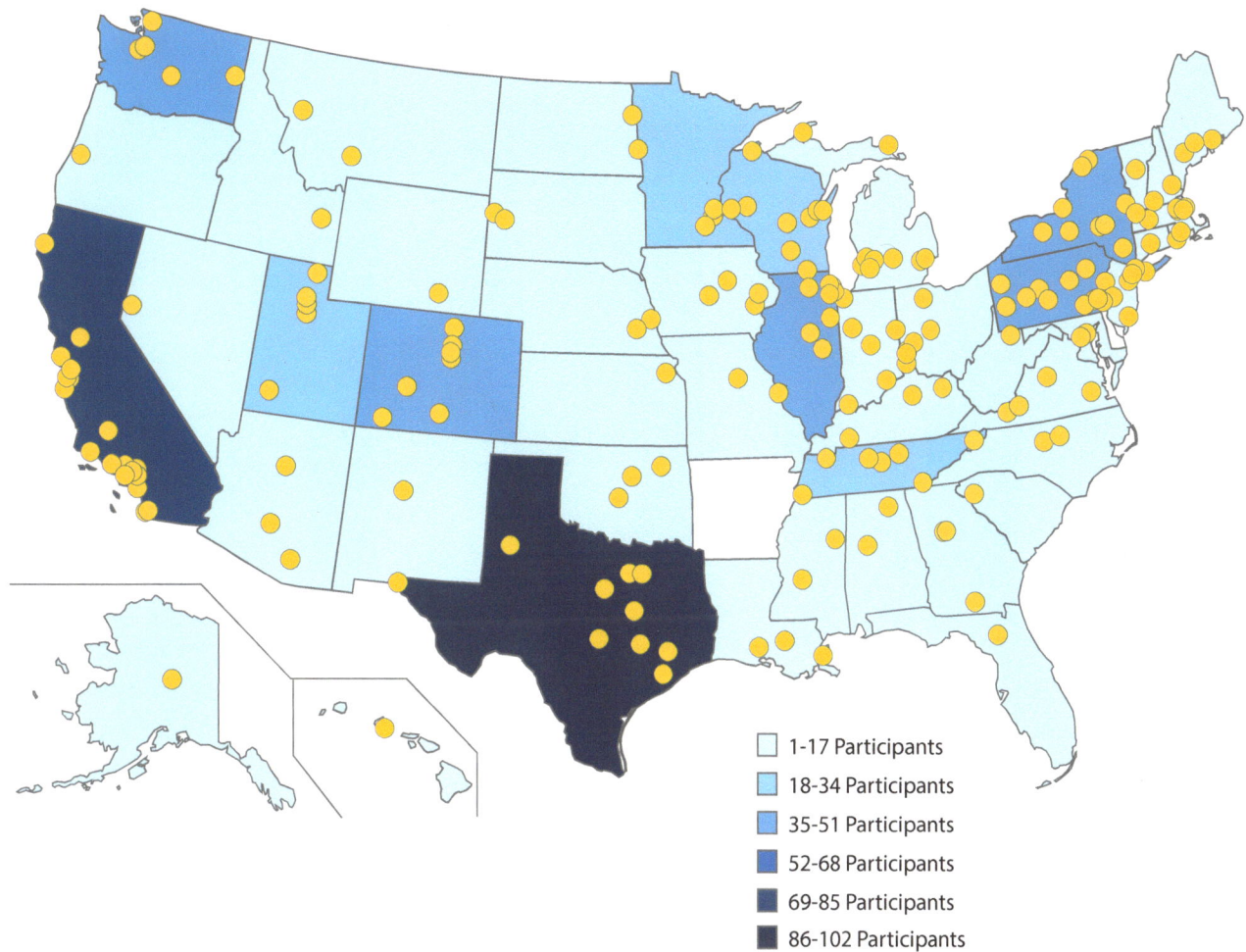

Legend:
- 1-17 Participants
- 18-34 Participants
- 35-51 Participants
- 52-68 Participants
- 69-85 Participants
- 86-102 Participants

The relative distribution by state of the universities and their graduating geoscience students across the United States that participated in the Exit Survey. *See Appendix I for list of departments

Degree recieved by participating graduates

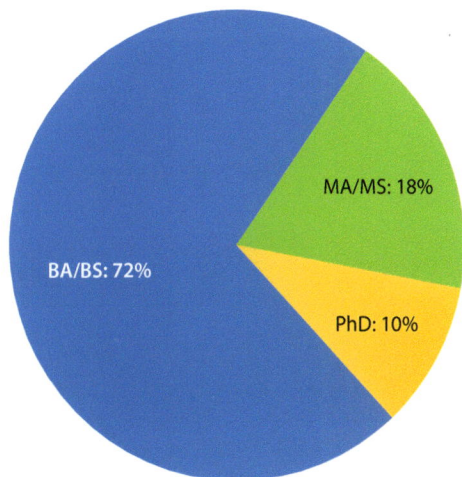

- BA/BS: 72%
- MA/MS: 18%
- PhD: 10%

Percentage of respondents within different classified institutions**

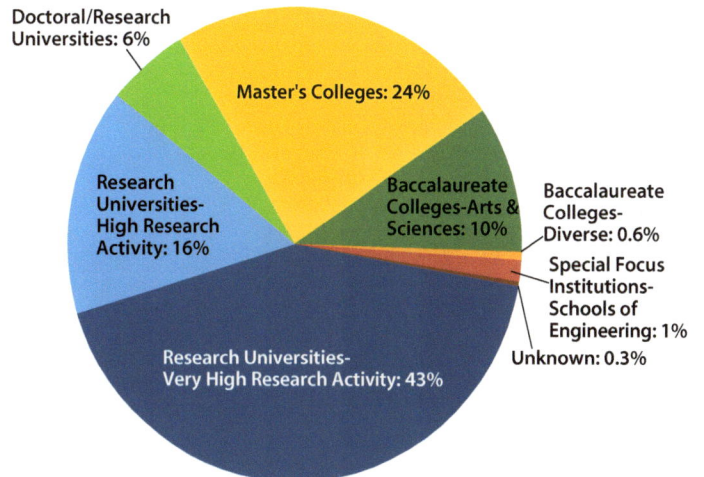

- Doctoral/Research Universities: 6%
- Master's Colleges: 24%
- Research Universities-High Research Activity: 16%
- Baccalaureate Colleges-Arts & Sciences: 10%
- Baccalaureate Colleges-Diverse: 0.6%
- Special Focus Institutions-Schools of Engineering: 1%
- Unknown: 0.3%
- Research Universities-Very High Research Activity: 43%

**See Appendix II for definitions of the Carnegie University Classification System

Gender breakdown of graduates

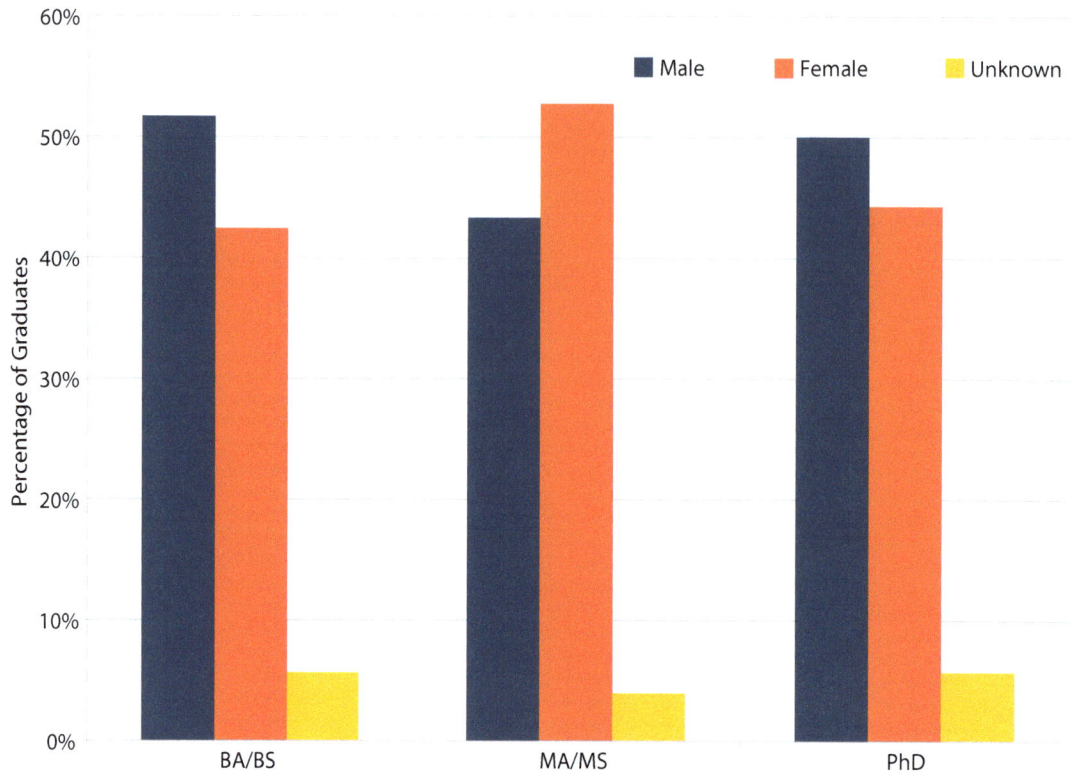

Age distribution of graduates

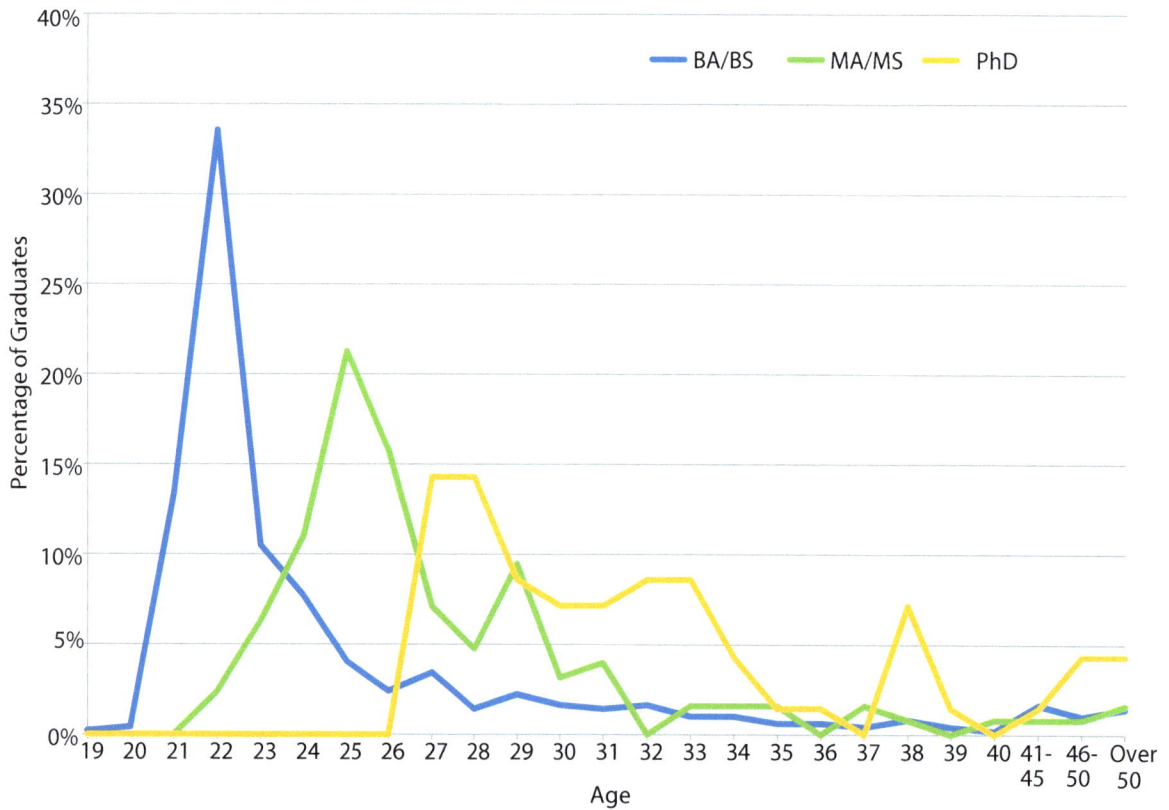

Citizenship of graduating students

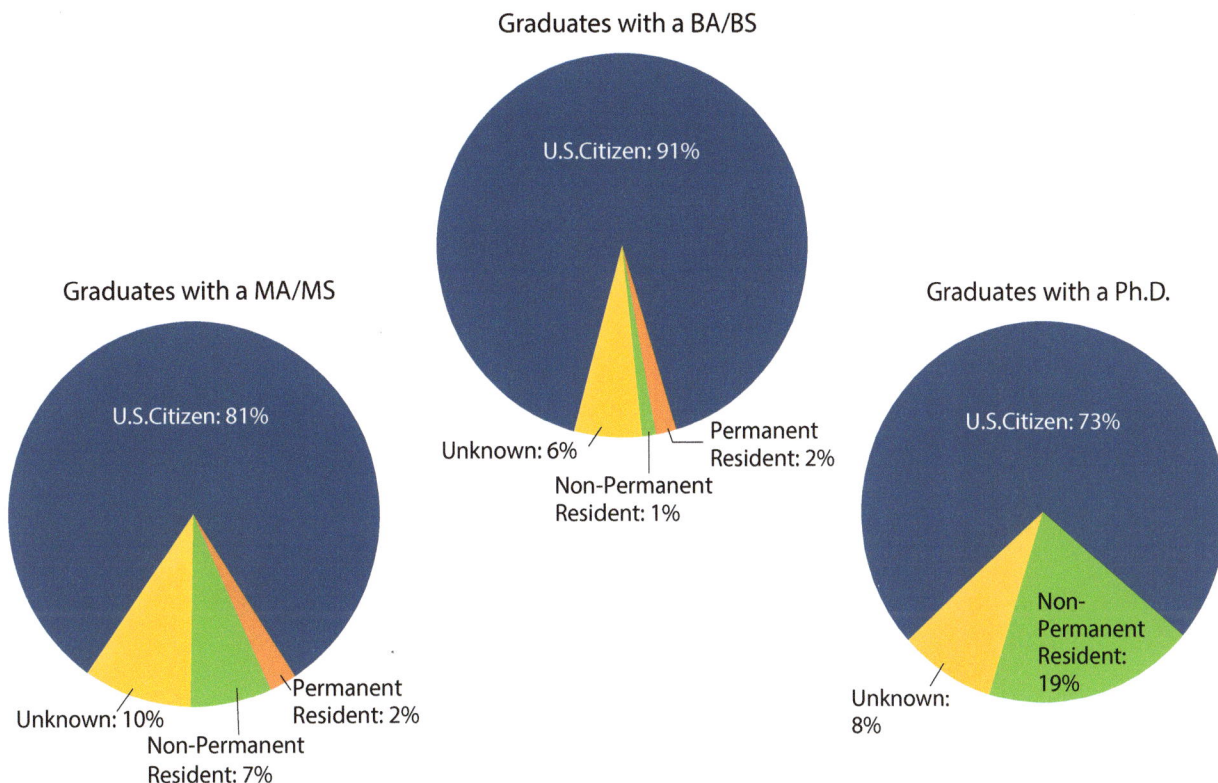

Graduates with a BA/BS

U.S.Citizen: 91%

Unknown: 6%

Non-Permanent Resident: 1%

Permanent Resident: 2%

Graduates with a MA/MS

U.S.Citizen: 81%

Unknown: 10%

Non-Permanent Resident: 7%

Permanent Resident: 2%

Graduates with a Ph.D.

U.S.Citizen: 73%

Non-Permanent Resident: 19%

Unknown: 8%

Race/ethnicity of graduating students

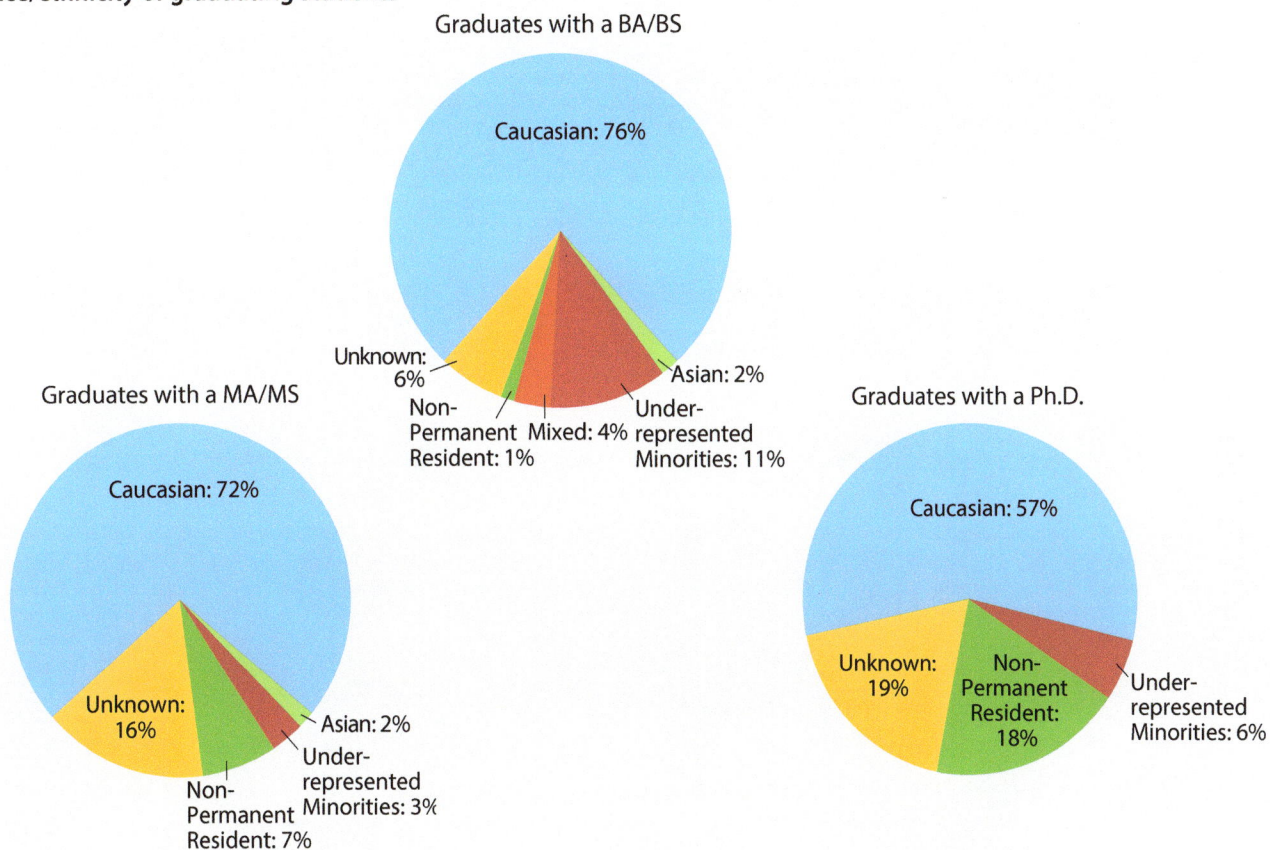

Graduates with a BA/BS

Caucasian: 76%

Unknown: 6%

Non-Permanent Resident: 1%

Mixed: 4%

Under-represented Minorities: 11%

Asian: 2%

Graduates with a MA/MS

Caucasian: 72%

Unknown: 16%

Non-Permanent Resident: 7%

Under-represented Minorities: 3%

Asian: 2%

Graduates with a Ph.D.

Caucasian: 57%

Unknown: 19%

Non-Permanent Resident: 18%

Under-represented Minorities: 6%

Highest education level of a parent/guardian of graduates

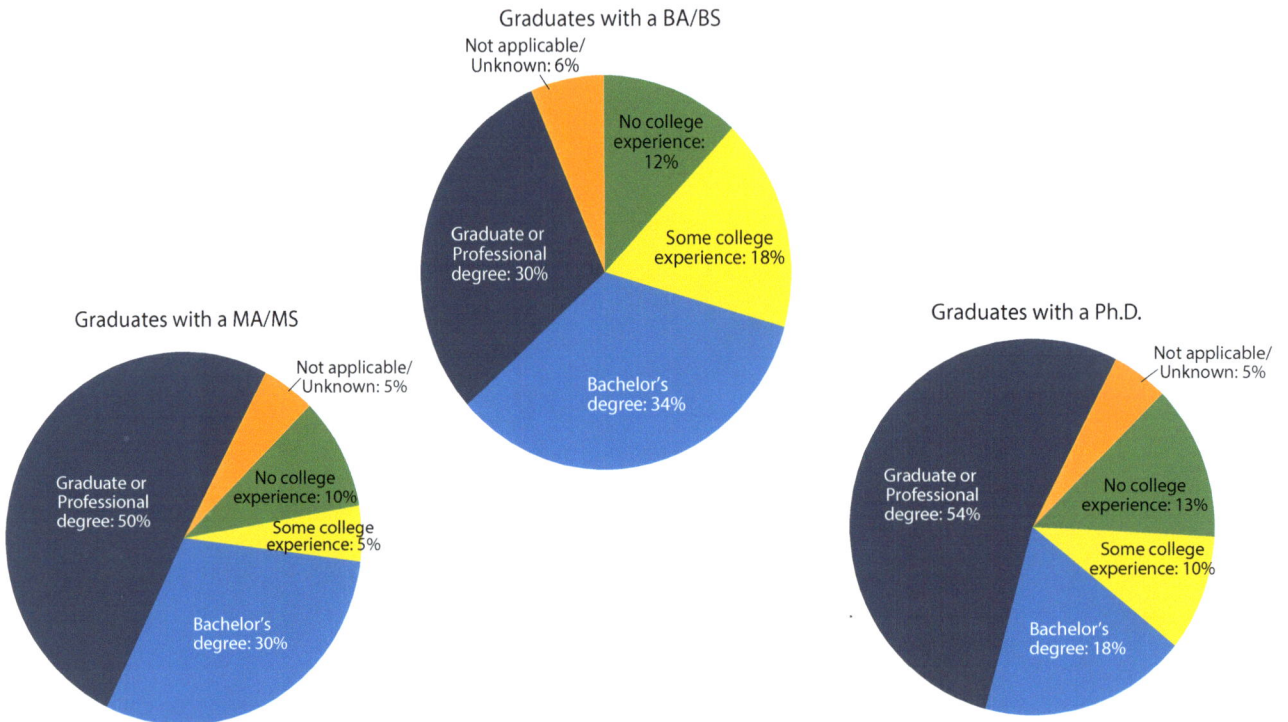

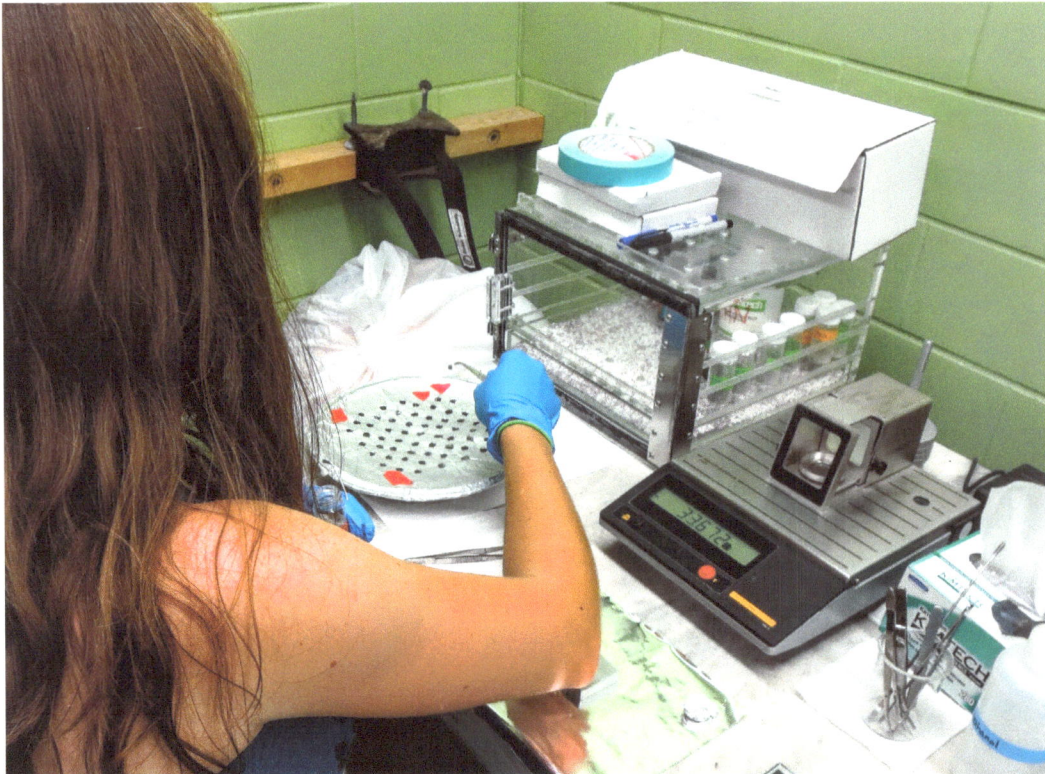

Graduates with a BA/BS

- Not applicable/Unknown: 6%
- No college experience: 12%
- Some college experience: 18%
- Bachelor's degree: 34%
- Graduate or Professional degree: 30%

Graduates with a MA/MS

- Not applicable/Unknown: 5%
- No college experience: 10%
- Some college experience: 5%
- Bachelor's degree: 30%
- Graduate or Professional degree: 50%

Graduates with a Ph.D.

- Not applicable/Unknown: 5%
- No college experience: 13%
- Some college experience: 10%
- Bachelor's degree: 18%
- Graduate or Professional degree: 54%

Photo by Sierra Melton from AGI's 2015 Life in the Field contest.

Port Aransas, Texas. Weighing out sediment samples to run through the elemental analyzer in order to determine organic carbon and nitrogen contents.

Photo by Dan Scott from AGI's 2015 Life in the Field contest.

Collecting soil cores to examine the controls on carbon storage in subalpine lake deltas. Location: Rawah Lake # 2, Rawah Wilderness, Colorado, USA.

Quantitative Skills and Geoscience Background of the Graduating Students

This section examines graduates' educational background, such as quantitative rigor, the role of K–12 experiences, and the importance of two-year colleges.

The students were asked to select all of the quantitative courses they have taken at a two-year or four-year institution. As in 2013 and 2014, most bachelor's graduates took Calculus II. Over half of the master's and doctoral graduates have taken up to Calculus III. A smaller percentage of bachelor's and master's graduates took higher level quantitative courses, and this largely represents generally the same cohort of students taking those higher level classes. A drop in the percentages of graduates at all degree levels taking Statistics was seen in 2015 compared to previous years. While over half of graduates at all levels indicate taking statistics, this still presents concerns about the ability of graduates at all levels to effectively read and interpret results using statistical methods and their skill in risk analysis. However, more students in 2015 compared to 2014 indicated having taken Differential Equations. When looking at the classifications of the institutions these students attended, it is clear that most of the students that took quantitative classes beyond Calculus II attended schools classified as high and very high research institutions. There does not appear to be a gender difference in the quantitative courses taken by recent geoscience graduates.

Along with identifying the quantitative courses taken, students were asked which additional core physical science courses they took during their postsecondary education. There was a slight decrease in the percentage of master's and doctoral graduates that indicated taking chemistry, but this may be due to a misunderstanding of the question. The question is meant to ascertain the percentage of graduates that have taken at least one course in chemistry or physics at some point during their higher education including prior degrees. There was an increase in the percentage of graduates that indicated taking calculus-based physics and a decrease in the percentage of graduates that took algebra-based physics indicating more confidence in the application of their quantitative skills.

Students were asked if they took an earth science course in high school and if they attended a two-year college for at least a semester before receiving a degree. Consistently from 2013–2015, the data from this survey has indicated the importance of the introduction of earth science in high school with approximately 50 percent of geoscience graduates taking at least one course before completing their high school diploma. While these courses may or may not be the reason a student majors in the geosciences, it does create an interest in the subject area and possibly a level of comfort in taking an introductory geoscience course during their first or second year in college. Each year, there has been an increase in the percentage of bachelor's graduates that spent at least a semester at a two-year college before completing the bachelor's degree. In 2015, that percentage rose to 33 percent of bachelor's graduates. Two-year colleges are becoming a viable and necessary option for many students to begin their post-secondary education, and this data supports the need for more collaborations and agreements between institutions to increase the ease of that transition.

Photo by Rachel Hatch from AGI's 2015 Life in the Field contest.

Ground Penetrating Radar survey at Lone Star Geyser, Yellowstone National Park.

Quantitative skills and knowledge gained while working towards degree

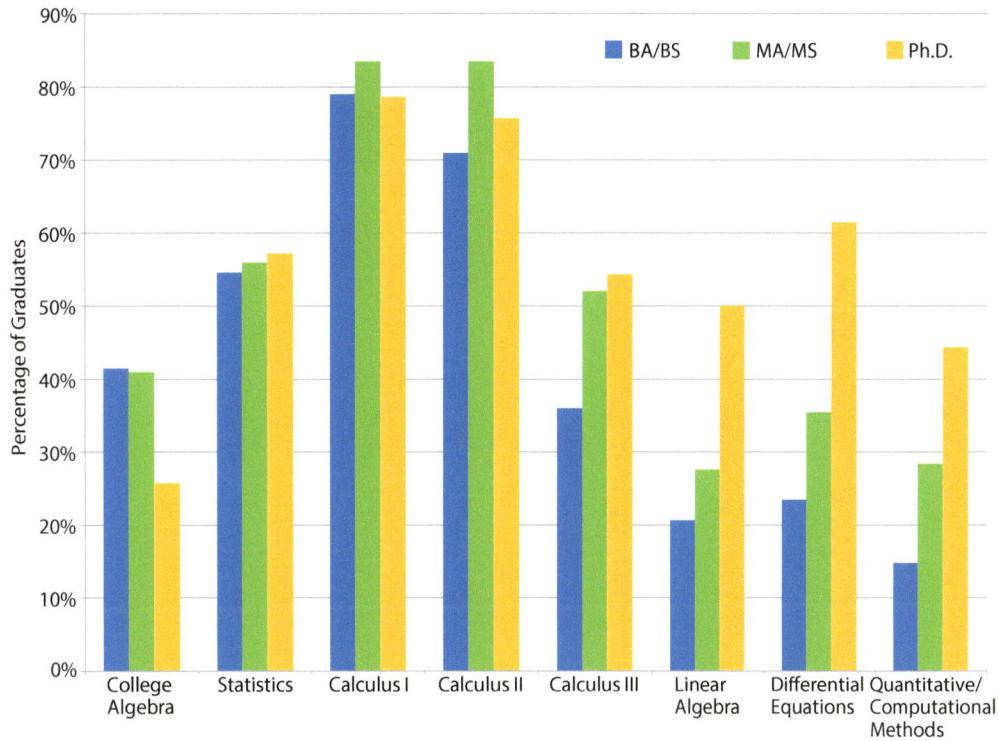

Quantitative skills and knowledge gained by graduates based on university classification**

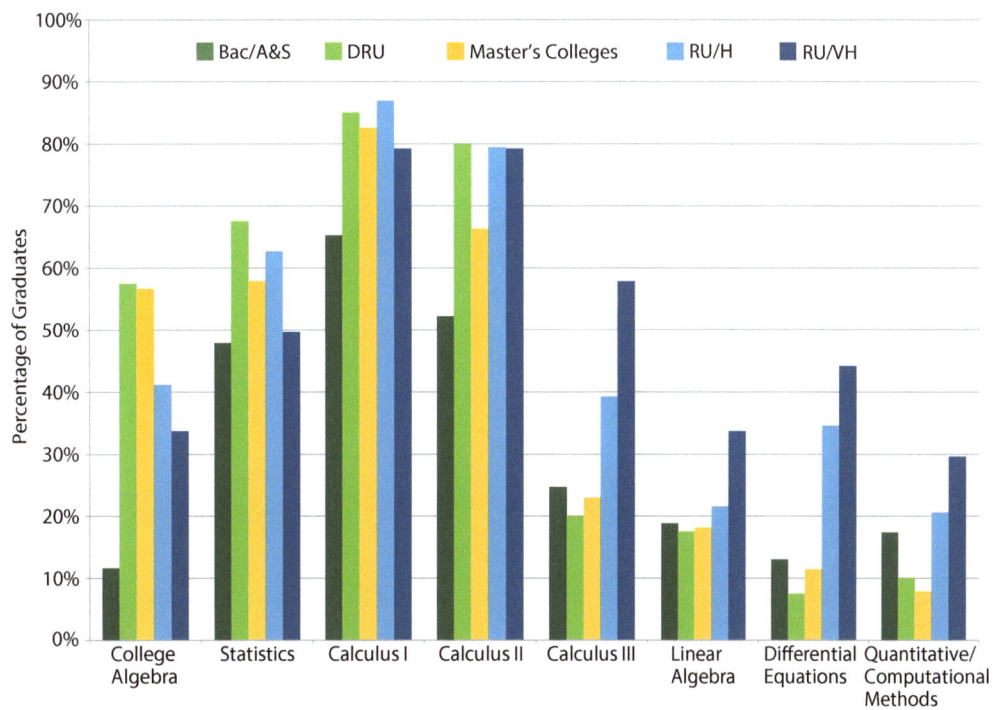

**See Appendix II for definitions of the Carnegie University Classification System

Quantitative skills and knowledge gained while working towards degree by gender

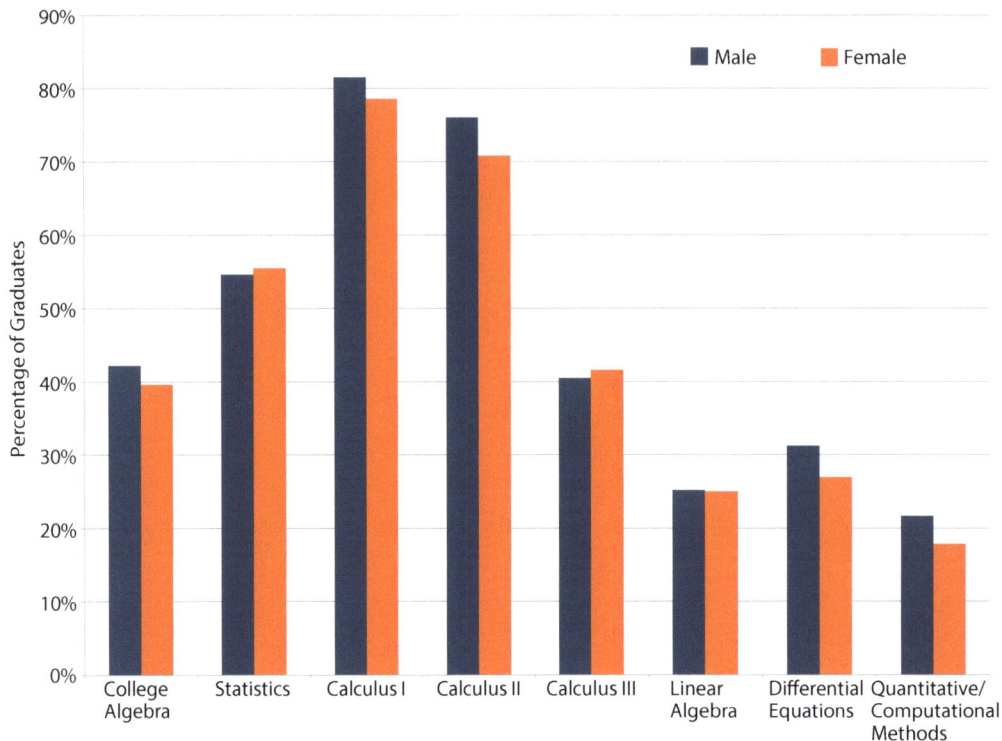

Percentage of graduates taking supplemental science courses

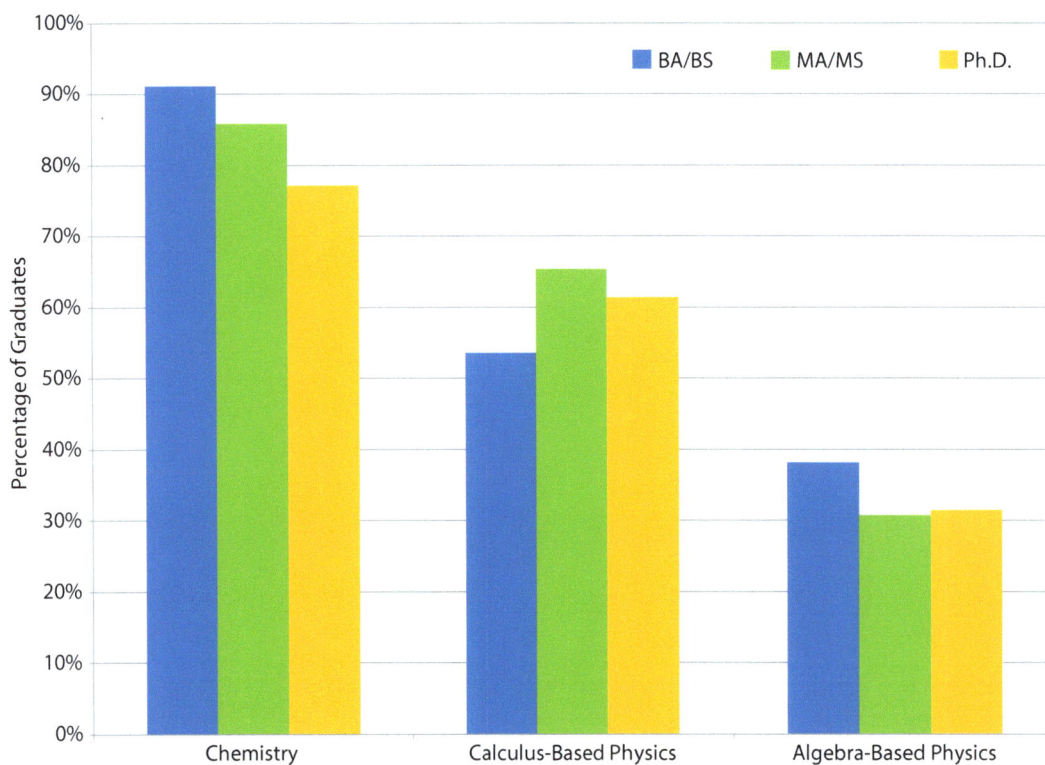

Graduates who took an earth science course in high school

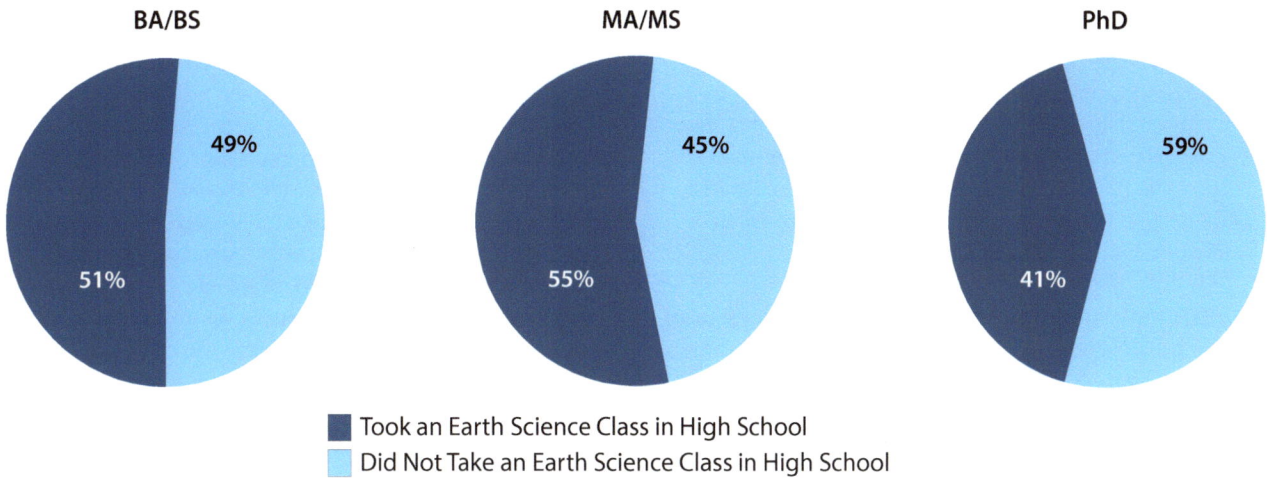

BA/BS	MA/MS	PhD
49% / 51%	45% / 55%	59% / 41%

■ Took an Earth Science Class in High School
■ Did Not Take an Earth Science Class in High School

Graduates who attended a two-year college for at least 1 semester and took a geoscience course

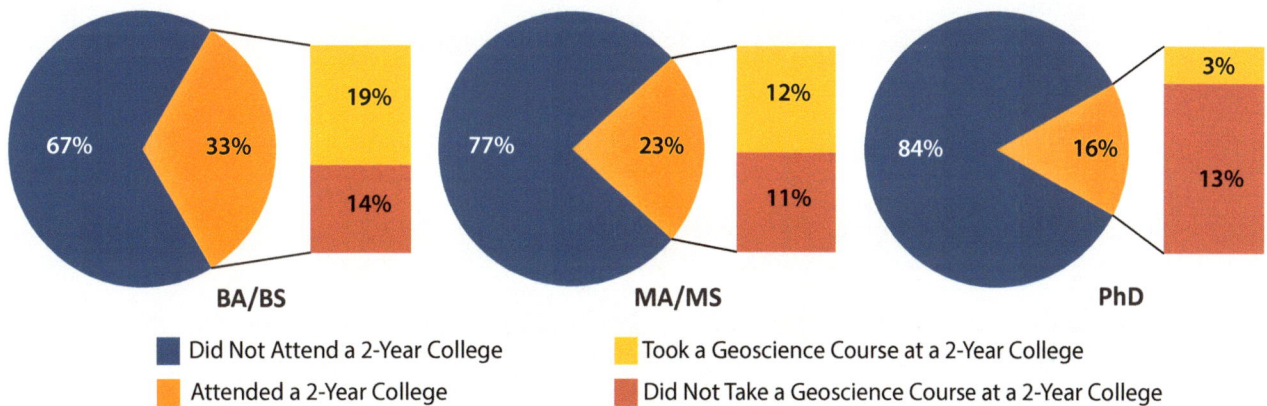

BA/BS	MA/MS	PhD
67% / 33% → 19% / 14%	77% / 23% → 12% / 11%	84% / 16% → 3% / 13%

■ Did Not Attend a 2-Year College   ■ Took a Geoscience Course at a 2-Year College
■ Attended a 2-Year College   ■ Did Not Take a Geoscience Course at a 2-Year College

Choosing Geoscience as a Major

Graduates were asked which geoscience field they were getting a degree in, as well as the fields associated with any other postsecondary degrees. The chosen degree fields demonstrate the variety of disciplines related to the geosciences. Geology continues to be the most popular degree among undergraduates with students specializing in different fields more often in graduate school.

In 2015, the majority of graduates at the bachelor's and master's levels chose to major in the geosciences at some point during their undergraduate educations. This trend has also been seen in previous years, which highlights the importance of the undergraduate geoscience courses to recruit majors. Most doctoral graduates in 2015 indicated choosing to major in the geosciences either after receiving an undergraduate degree or beginning their collegiate education. The graduates were asked to explain their reasoning for majoring in the geosciences. As in 2014, the majority of graduates at all levels indicated the intellectual engagement of the geosciences as the reason for choosing their major. These comments often included reasons related to the interdisciplinary nature of the geosciences, the applicability of math and other sciences to earth processes, the skills development and problem solving required for the geosciences, and the passion for particular fields within the geosciences. Beyond the intellectual engagement of the geosciences, bachelor's and master's graduates also commented more often on the career opportunities in the geosciences, desire to understand the societal impacts of the geosciences, and the importance of the department culture and/or particular faculty encouraging them to their major. Doctoral graduates commented more often on career opportunities, the desire to understand the societal impacts of the geosciences, and the love of spending time outdoors. Other comments by geoscience graduates in 2015 included the enjoyment of their introductory courses, family influences toward the geosciences, a long time childhood interest in the field, and the enjoyment of field work, trips, and research.

When students decide to major in the geosciences

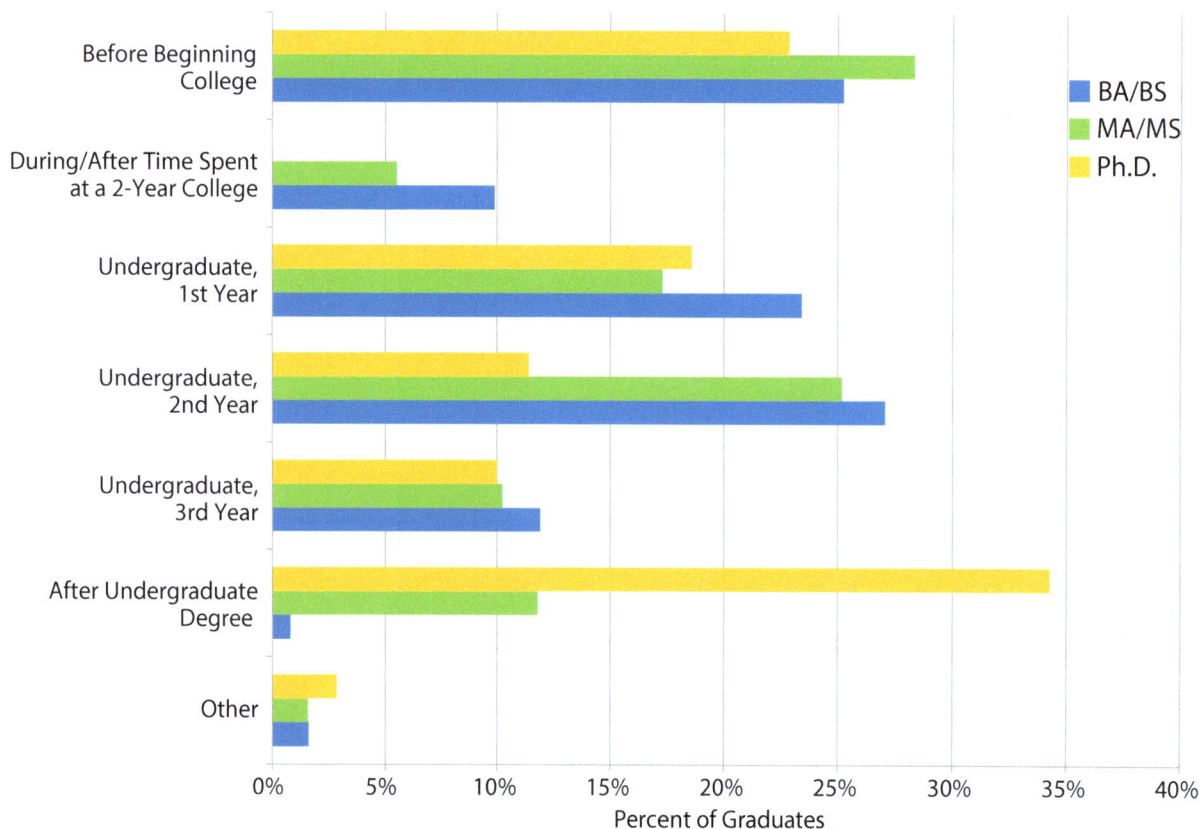

Chosen geoscience degree fields

Bachelor's Degree Graduates

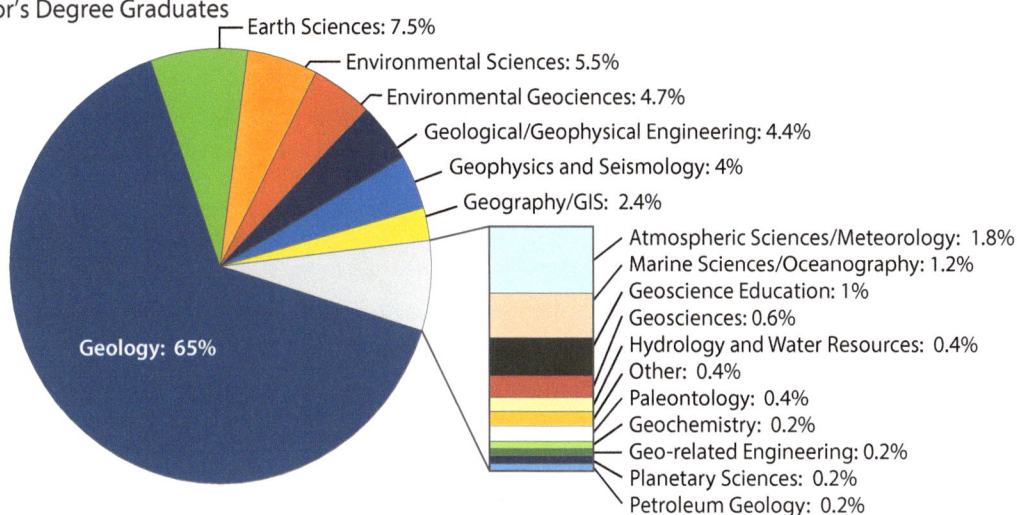

- Earth Sciences: 7.5%
- Environmental Sciences: 5.5%
- Environmental Geociences: 4.7%
- Geological/Geophysical Engineering: 4.4%
- Geophysics and Seismology: 4%
- Geography/GIS: 2.4%
- Atmospheric Sciences/Meteorology: 1.8%
- Marine Sciences/Oceanography: 1.2%
- Geoscience Education: 1%
- Geosciences: 0.6%
- Hydrology and Water Resources: 0.4%
- Other: 0.4%
- Paleontology: 0.4%
- Geochemistry: 0.2%
- Geo-related Engineering: 0.2%
- Planetary Sciences: 0.2%
- Petroleum Geology: 0.2%
- Geology: 65%

Master's Degree Graduates

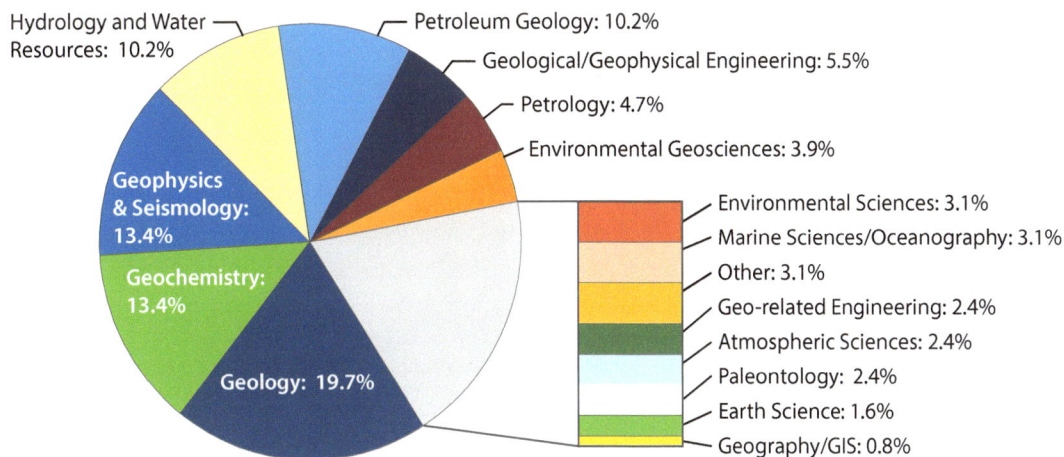

- Hydrology and Water Resources: 10.2%
- Petroleum Geology: 10.2%
- Geological/Geophysical Engineering: 5.5%
- Petrology: 4.7%
- Environmental Geosciences: 3.9%
- Geophysics & Seismology: 13.4%
- Geochemistry: 13.4%
- Geology: 19.7%
- Environmental Sciences: 3.1%
- Marine Sciences/Oceanography: 3.1%
- Other: 3.1%
- Geo-related Engineering: 2.4%
- Atmospheric Sciences: 2.4%
- Paleontology: 2.4%
- Earth Science: 1.6%
- Geography/GIS: 0.8%

Doctoral Degree Graduates

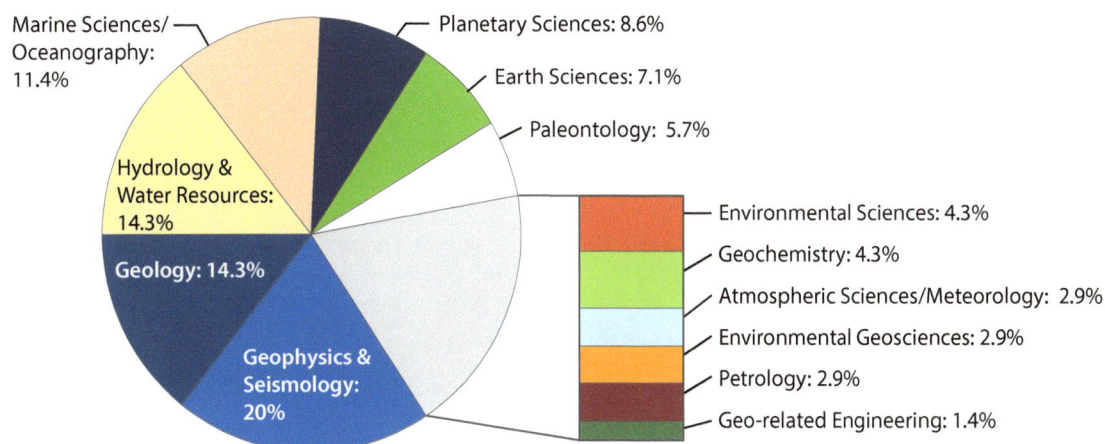

- Marine Sciences/Oceanography: 11.4%
- Planetary Sciences: 8.6%
- Earth Sciences: 7.1%
- Paleontology: 5.7%
- Hydrology & Water Resources: 14.3%
- Geology: 14.3%
- Geophysics & Seismology: 20%
- Environmental Sciences: 4.3%
- Geochemistry: 4.3%
- Atmospheric Sciences/Meteorology: 2.9%
- Environmental Geosciences: 2.9%
- Petrology: 2.9%
- Geo-related Engineering: 1.4%

Ancillary Factors Supporting the Degree

Graduates were asked about their experiences while working towards their degree. In 2015, as in previous years, the majority of bachelor's and doctoral graduates did not hold an internship during their postsecondary education. However, this year did see an increase of 14 percent of master's graduates participating in one or more internships before completing their degree. It is possible that the 2015 master's graduates understood the importance of internships to their professional career after graduation. Two questions were added to the 2015 survey asking how many internship applications were submitted by recent graduates and how they found the internship opportunities. Approximately 26 percent of bachelor's graduates and 11 percent of master's graduates that did not participate in an internship did apply for at least one opportunity. Most of the recent graduates recognize the importance of internships for their professional development, but there may not be enough opportunities available for interested students. The geoscience community may need to consider other ways to provide this type of professional development to current geoscience students.

It is clear that these recent graduates, particularly the master's graduates, used whatever resources were at their disposal to find internship opportunities. Bachelor's graduates tended to focus on faculty referrals and internet searches to find these opportunities, and doctoral graduates tended to use their professional networks for this search. Of those students that participated in an internship, 80 percent percent of bachelor's graduates, 88 percent of master's graduates, and 73 percent of doctoral graduates rated their internships as "very important" to their academic and professional development.

When asked about the types of financial aid used to fund their education, 76 percent of bachelor's graduates, 89 percent of master's graduates, and 93 percent of doctoral graduates indicating using some form of financial aid while working towards their degree. As in past years, higher percentages of bachelor's graduates tended to use student loans, work study appointments, and federal grants to fund their education compared to students working toward a graduate degree. Geoscience graduates students tend to depend mostly on research and teaching assistantships to fund their education, but there are some students that still use student loans to pay for a part of their graduate education.

Graduates were also asked about their involvement with geoscience membership organizations. AGI is a federation of 51 geoscience societies, including the American Geophysical Union, the American Institute of Professional Geologists, the Association for the Sciences of Limnology and Oceanography, the Geological Society of America, and the Society of Exploration Geophysicists. Professional societies can be useful tools for success as an early-career geoscientist. Considering these five societies helped distribute the survey in 2015 to their student membership, the percentages of participation as student members of a geoscience society was surprisingly low among bachelor's and doctoral graduates. There was a slight increase in the percentage of master's graduates indicating their membership to an AGI member society compared to 2014.

Photo by Rachel Hatch from AGI's 2015 Life in the Field contest.

Magnetometer survey at Mammoth Lakes, CA.

Number of internships held by graduating students

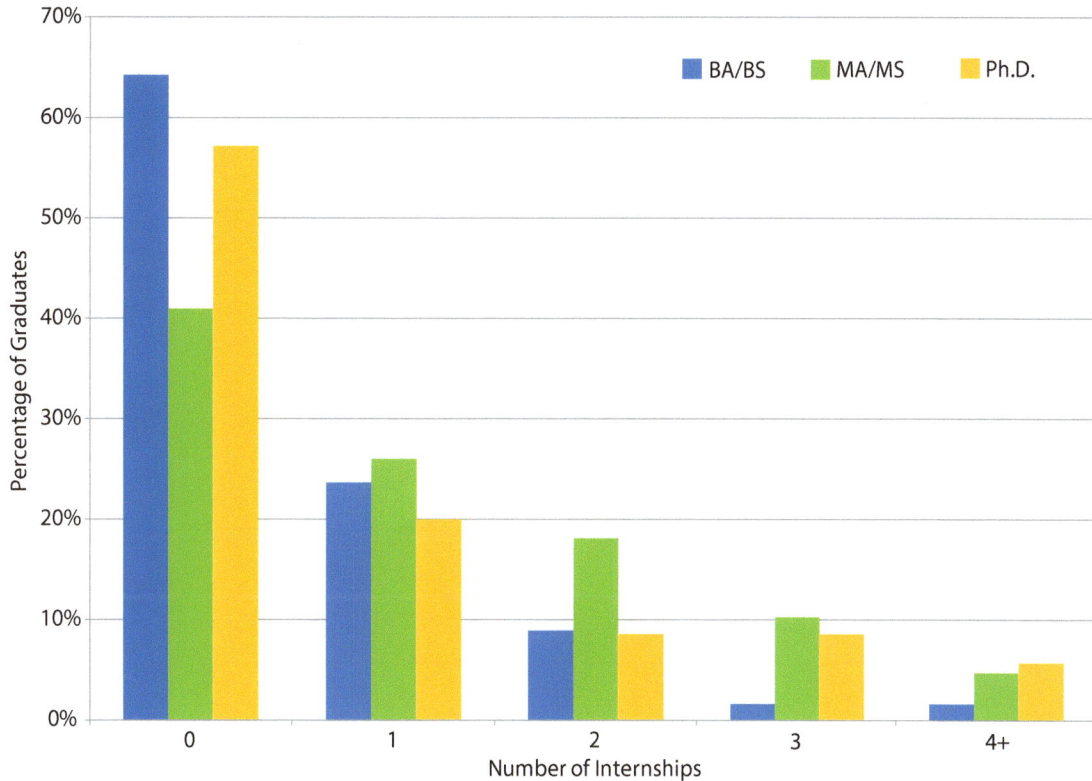

Internship applications completed by graduates

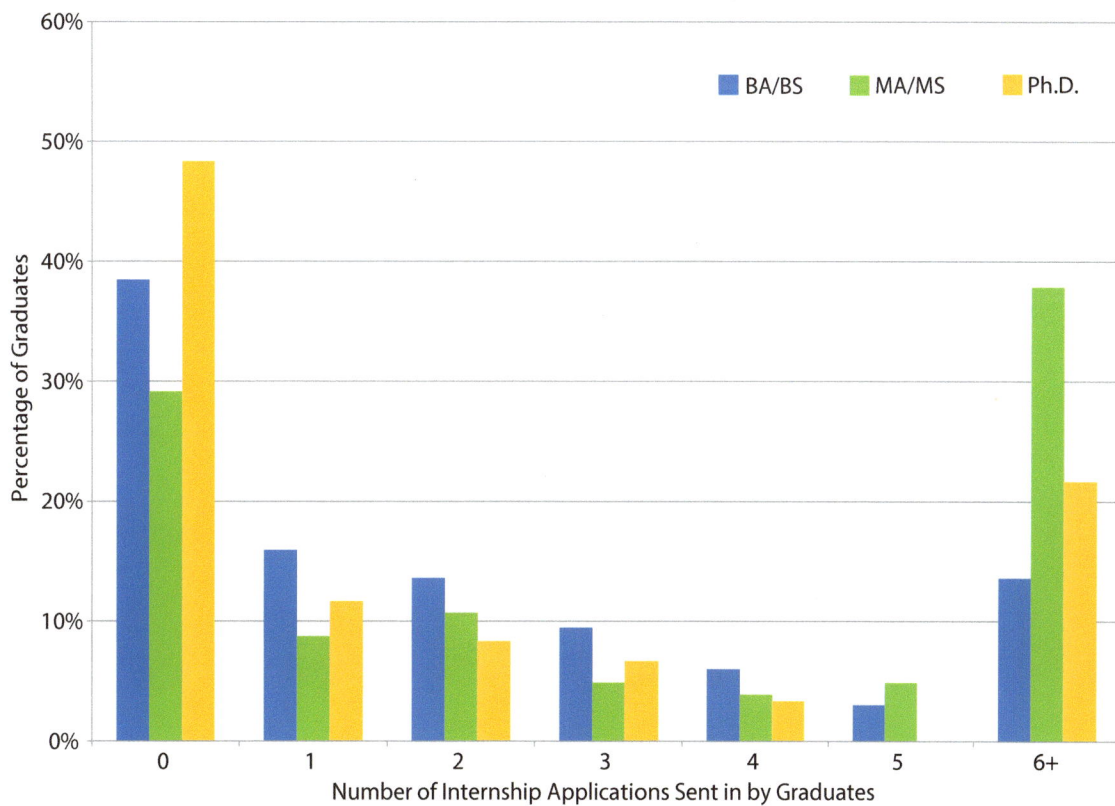

Resources used to find internship announcements

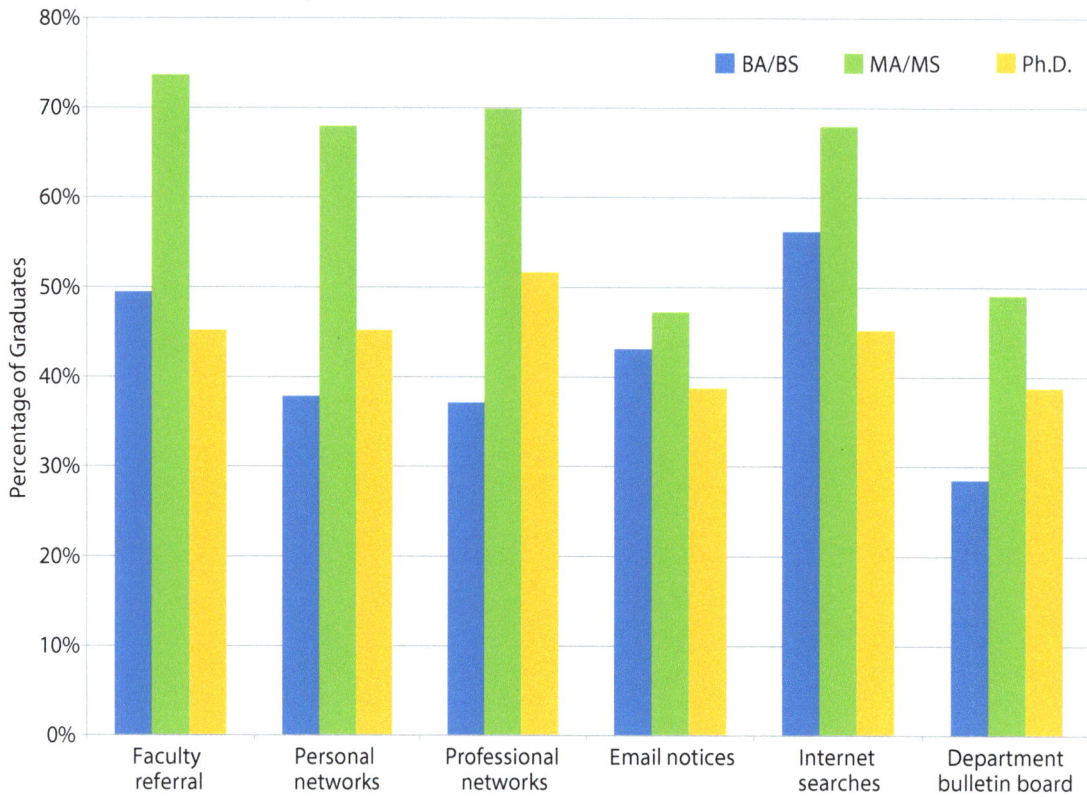

Types of financial aid used by graduating students while working towards a degree

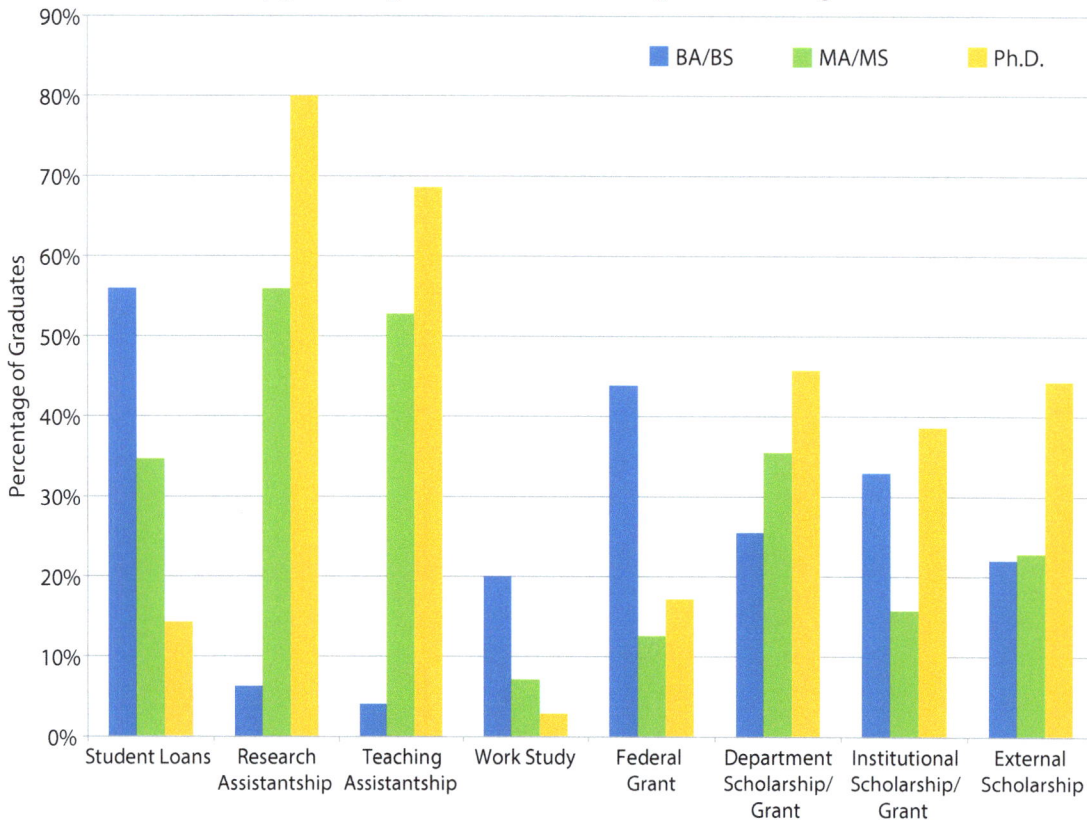

Participation in geoscience clubs

	BA/BS	MA/MS	Ph.D.
Associated with a geoscience-related club/organization	69%	80%	60%
Participated in department-level geoscience club	58%	56%	36%
Member of an AGI Member Society	21%	45%	27%
Member of an Honor Society	7%	8%	3%

Average GPA

	BA/BS	MA/MS	Ph.D.
Average years to degree completion	4.20	3.29	6.20
Average overall GPA	3.29	3.69	3.86
Average geoscience GPA	3.41	3.73	3.91

Photo by Filipe Pinto from AGI's 2015 Life in the Field contest.

Field Experiences

Clear definitions were set to distinguish between field camp, field courses, and field experiences. A field camp was defined as an academic program lasting four or more weeks that is primarily focused on field tools and methods. A field course was defined as a course with a field component primarily covering field methods and experimentation that utilized at least half of the total class time. A field experience was defined as any course that contained a field component, such as a field trip, field work, or other time in the field, that is not included in the definitions for field camp or field courses.

In 2015, only about 2 percent of geoscience graduates did not participate in a field experience of some kind while working towards their degree. In 2015, there was a 10 percent decrease in the percentage of doctoral graduates that participated in field camp during their postsecondary education compared to 2014, but the participation by bachelor's and master's graduates stayed relatively similar to the previous year. As in 2014, there is a gender difference in the field camp participation with approximately 7 percent more men attending field camp before graduation than women. When asked about the importance of these field experiences to the graduates' academic and professional development, field camps, field courses, and field experiences were all rated as "very important" by the majority of graduates. As in previous years, this rating was highest for all degree levels for field experiences, with 86 percent of bachelor's graduates, 80 percent of master's graduates, and 86 percent of doctoral graduates finding these experiences "very important" to their academic and professional development.

Graduates' participation in field experiences was also broken down by the Carnegie classification of the institution (see Appendix II) to see if the type of institution these students attended may have affected their access to field experiences. While field experiences are prevalent at all the participating universities, the baccalaureate colleges and the master's colleges continue to have the lowest participation rates in field camp amongst their graduates, with only a slight increase in the participation rate at baccalaureate colleges compared to 2014.

Student participation in field experiences based on university classification**

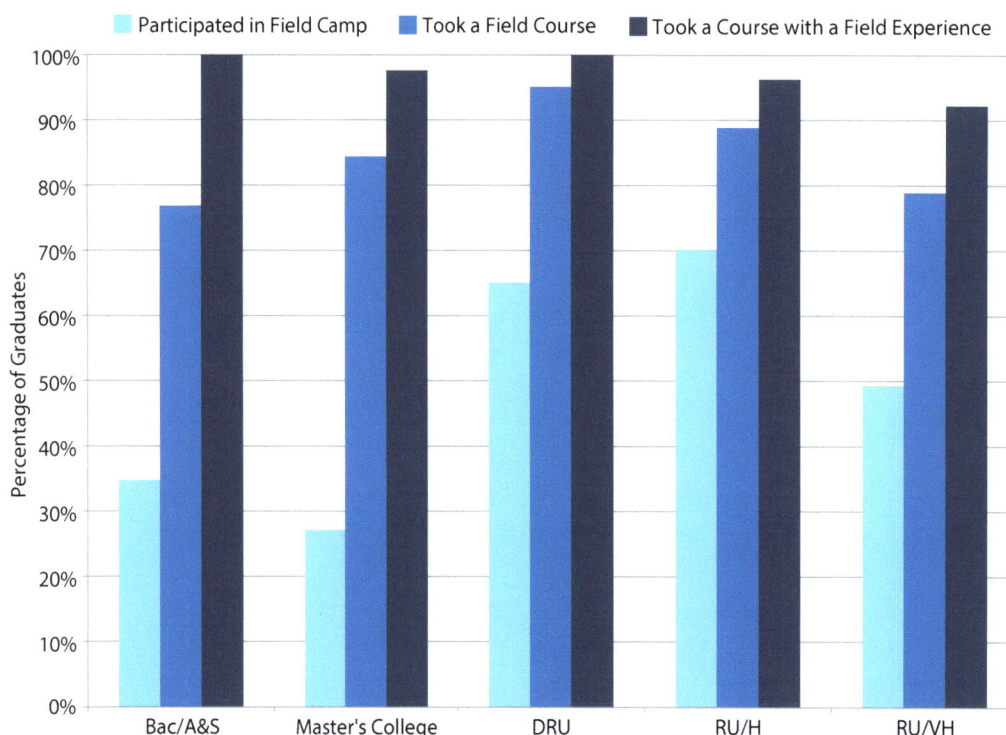

**See Appendix II for definitions of the Carnegie University Classification System

Graduating students who have participated in field camp

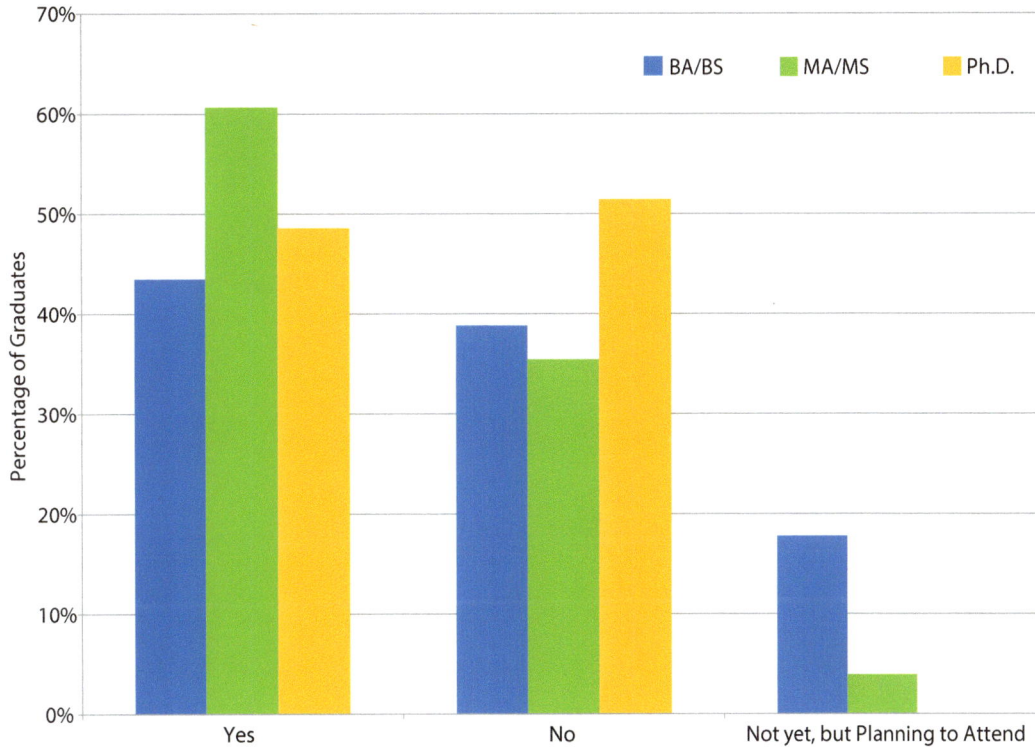

Graduating students who have participated in field camp by gender

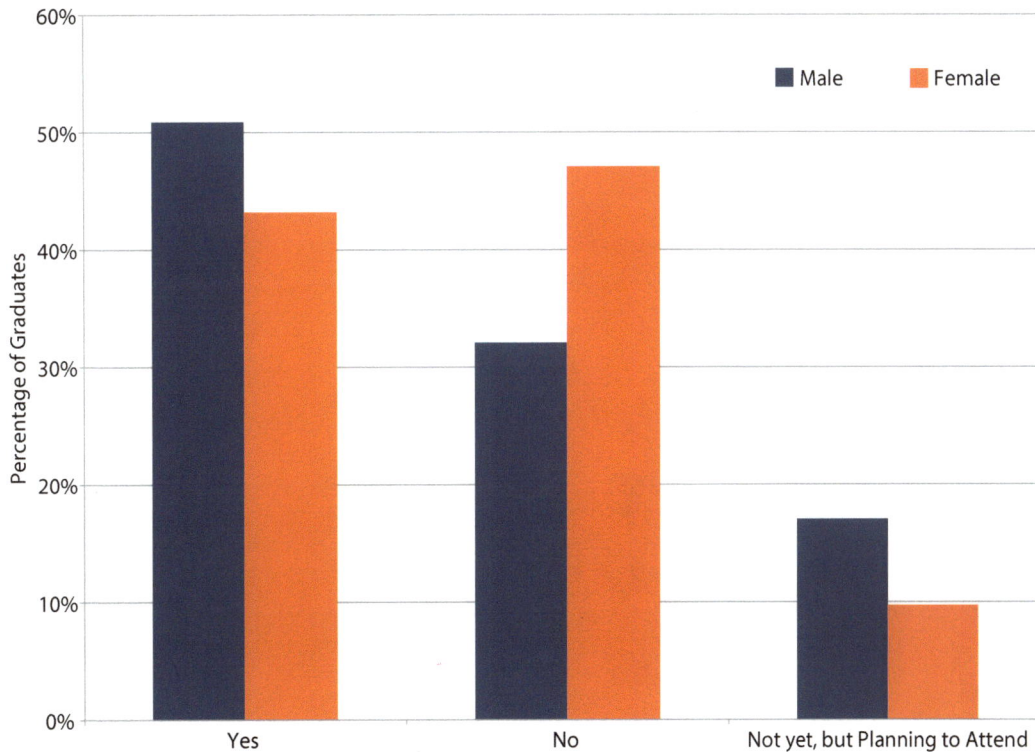

Number of field courses taken by graduates

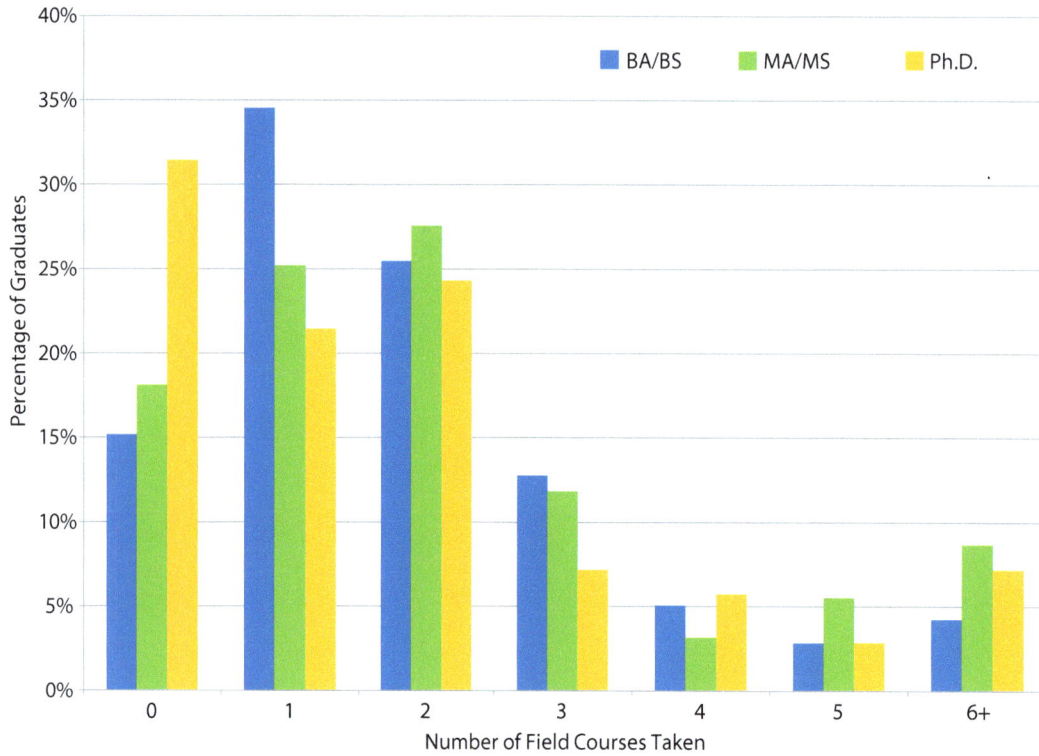

Courses taken with field experiences by graduates

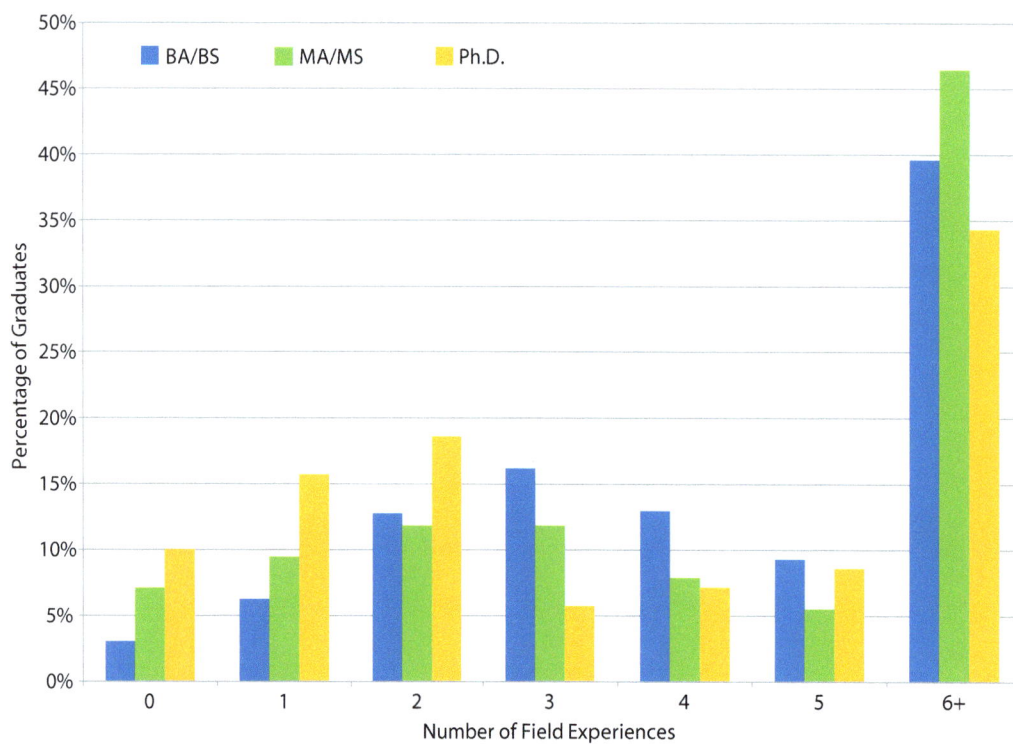

Photo by Hanif Jawid from AGI's 2015 Life in the Field contest.
Field data visualization at site office with groups, at location of Parwan province Afghanistan.

Research Experiences

The graduates were asked about their research experiences while working toward their degrees. If they indicated participation in at least one research experience, the graduates were then asked about their participation in faculty-directed research and self-directed research. If they indicated participation in self-directed research, they were asked to identify the basic research methodology used to conduct their research.

In 2015, there was a 7 percent decrease in the bachelor's graduates that participated in at least one research experience compared to 2014. Not surprisingly, graduates participated in more research experiences at the doctoral degree level than the bachelor's of master's degree levels with 47 percent of doctoral graduates indicating have six or more difference research experiences.

For 2015, geochemistry and planetary sciences were added as degree fields to the figure displaying the percentage of research methods utilized by the recent graduates. Literature-based research was not used as often by undergraduate and graduate students receiving their degree in 2015 compared to 2014, decreasing by 10 percent and 7 percent respectively. Field-based and lab-based research methods continue to be more often used by undergraduates compared to the other research methods, and graduate students in 2015 used computer-based methods a bit more often than the other research methods.

When asked about the importance of research experiences to the graduates' academic and professional development, 81 percent of bachelor's graduates, 89 percent of master's graduates, and 99 percent of doctoral graduates rated these experiences as "very important".

Research methods utilized by graduates in their self-directed research

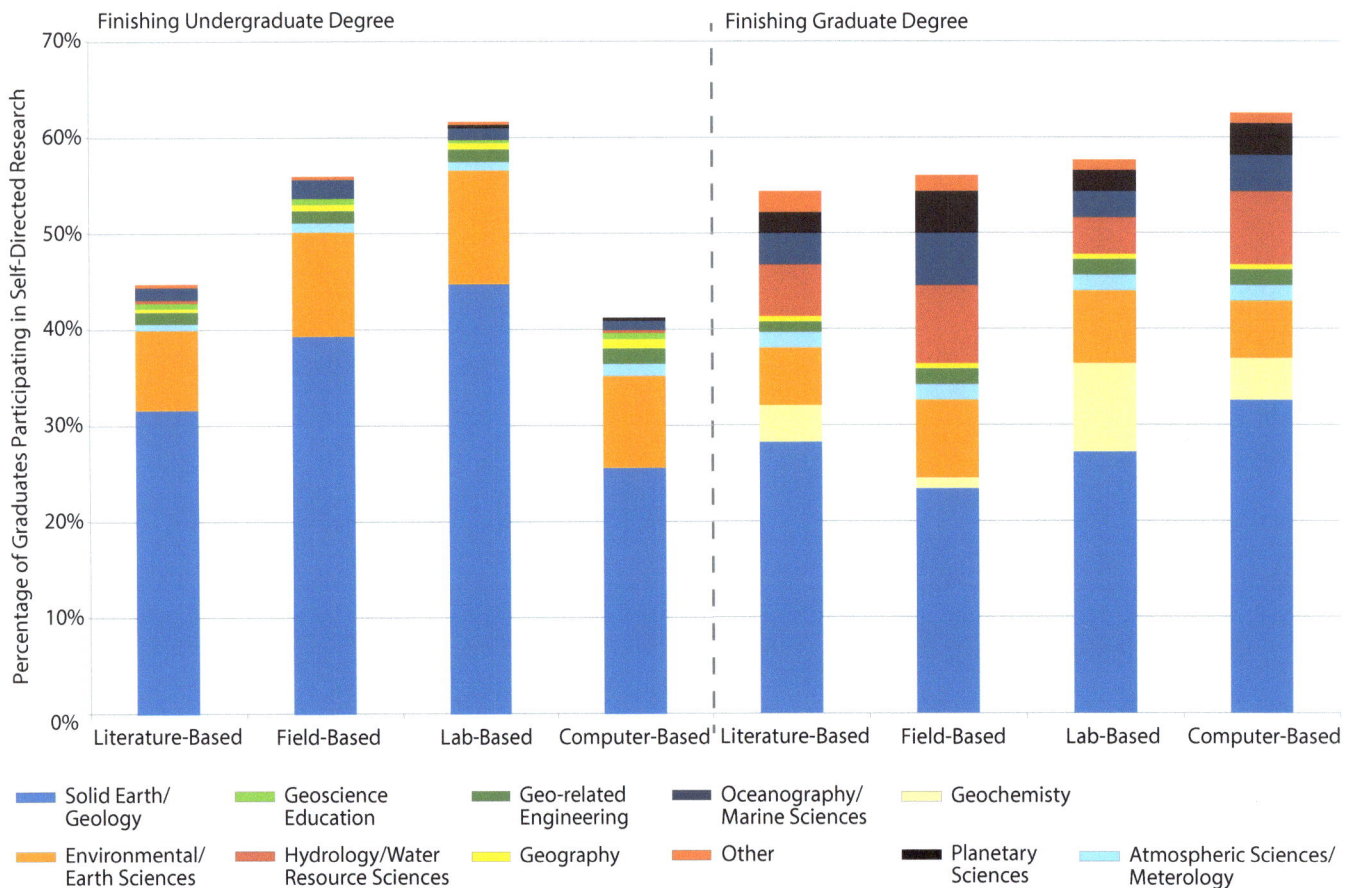

Legend:
- Solid Earth/Geology
- Geoscience Education
- Geo-related Engineering
- Oceanography/Marine Sciences
- Geochemisty
- Environmental/Earth Sciences
- Hydrology/Water Resource Sciences
- Geography
- Other
- Planetary Sciences
- Atmospheric Sciences/Meterology

Participation rates of graduates in research experiences

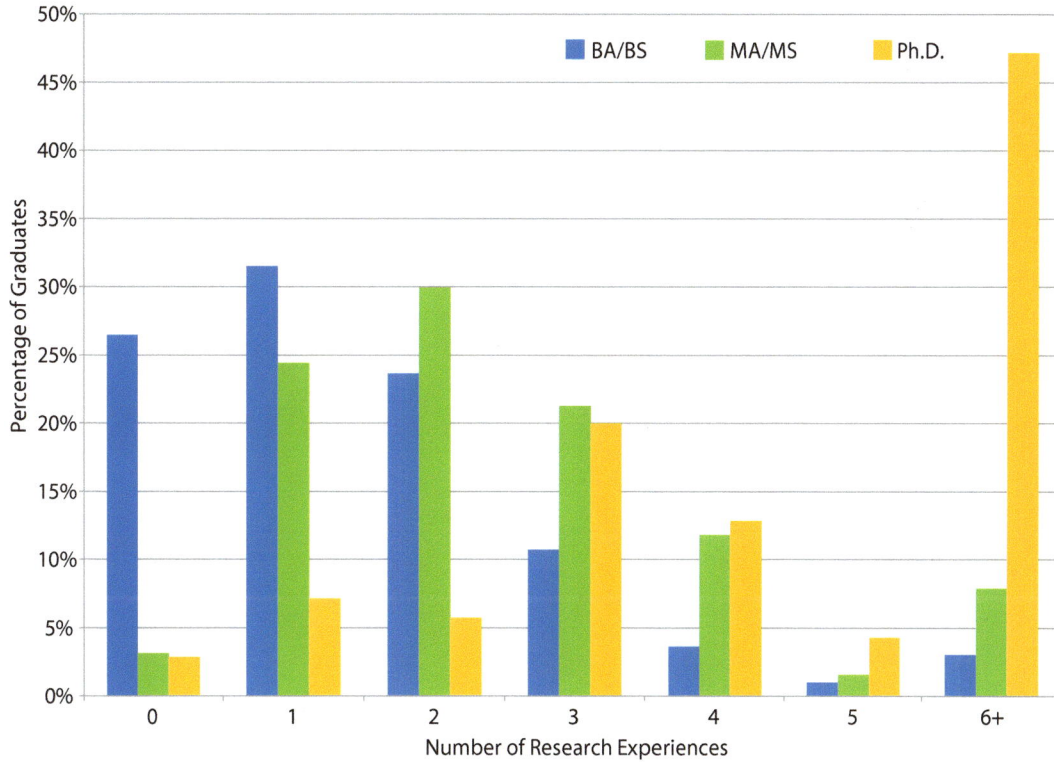

Bar chart with legend: BA/BS (blue), MA/MS (green), Ph.D. (yellow). Y-axis: Percentage of Graduates (0% to 50%). X-axis: Number of Research Experiences (0, 1, 2, 3, 4, 5, 6+).

Student participation in faculty-directed and self-directed research

	BA/BS	MA/MS	Ph.D.
Faculty-Directed Research	49%	78%	86%
Self-Directed Research	63%	91%	97%

Research methods utilized by graduates in their self-directed research by gender

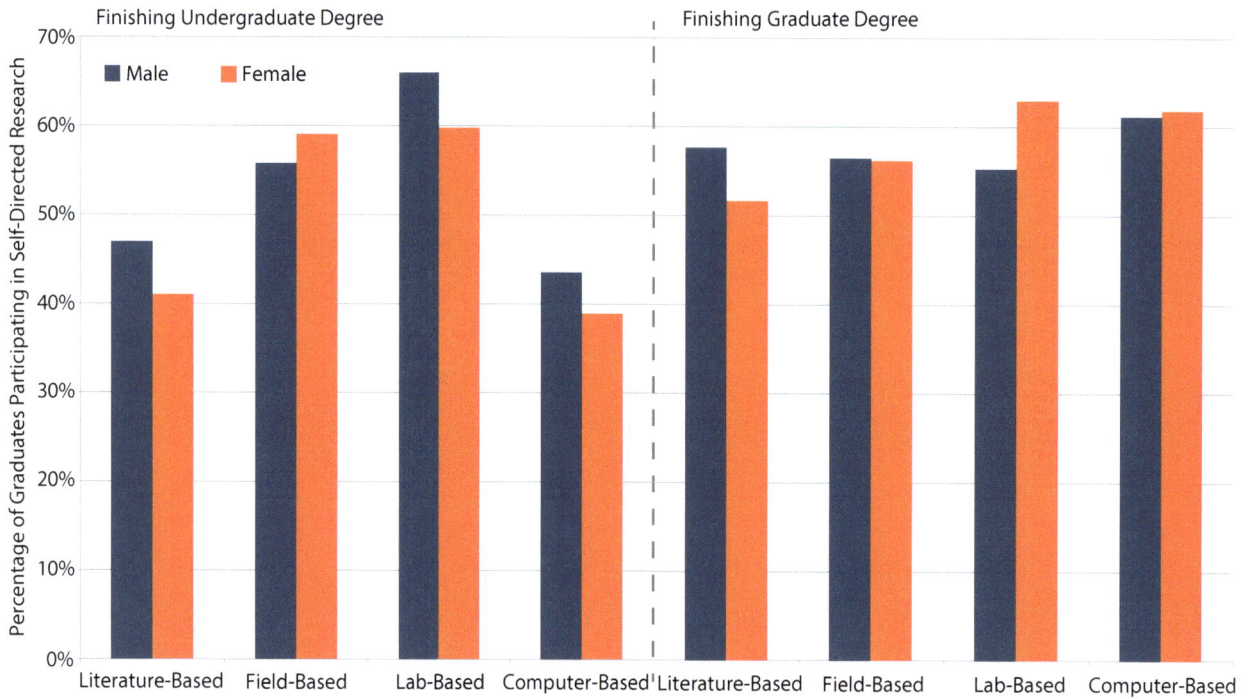

Finishing Undergraduate Degree | Finishing Graduate Degree

■ Male ■ Female

Y-axis: Percentage of Graduates Participating in Self-Directed Research

Student participation in research based on university classification**

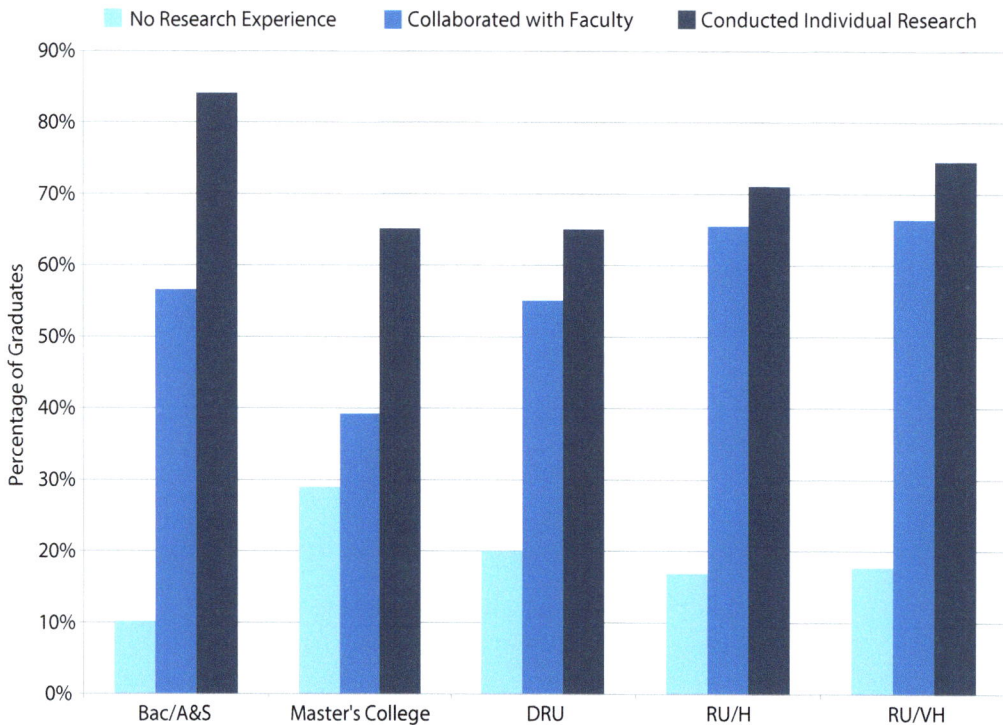

■ No Research Experience ■ Collaborated with Faculty ■ Conducted Individual Research

Y-axis: Percentage of Graduates

X-axis: Bac/A&S, Master's College, DRU, RU/H, RU/VH

**See Appendix II for definitions of the Carnegie University Classification System

Photo by Hannah Cunningham from AGI's 2015 Life in the Field contest.

Future Plans: Working Toward a Graduate Degree

The graduates were asked if they have immediate plans to continue their education. Those indicating plans for a graduate degree after graduation were then asked to share the degree they would pursue and the field of interest for the degree.

The percentage of bachelor's graduates planning to attend graduates school dropped from 42 percent in 2014 to 38 percent in 2015 — the same percentage of bachelor's graduates in 2013. The percentage of master's graduates planning to get another graduate degree also dropped from 26 percent in 2014 to 20 percent in 2015. Recently, departments have expressed concern with their ability to take on more graduate students because they have reached their capacity for students. The drop in percentage of recent graduates planning to attend graduate school may have been in response to the competitiveness in recent

years getting into graduate programs in the U.S. There is also the added pressure of student debt among recent graduates, which might encourage many of them to try and find a job and waiting a few years before incurring further educational debt.

New bachelor's graduates showed a varied array of intended future graduate degree fields. The fields that fell into the "Other" category included materials science and natural resource science. The fields that fell in the "Other" category among graduates planning on a second graduate degree included agricultural engineering and a few undecided fields.

In 2015, there was an increase of 13 percent of bachelor's graduates and an increase of 39 percent of master's graduates planning to pursue doctoral degrees compared to 2014.

Students planning to attend graduate school after graduation

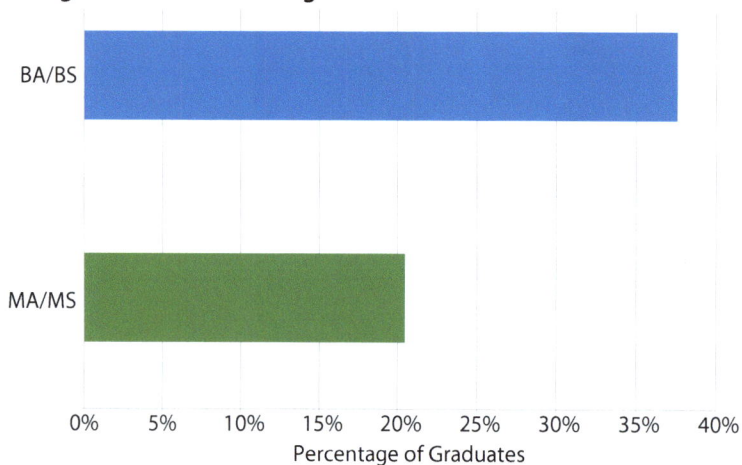

Students planning to attend graduate school after graduation by gender

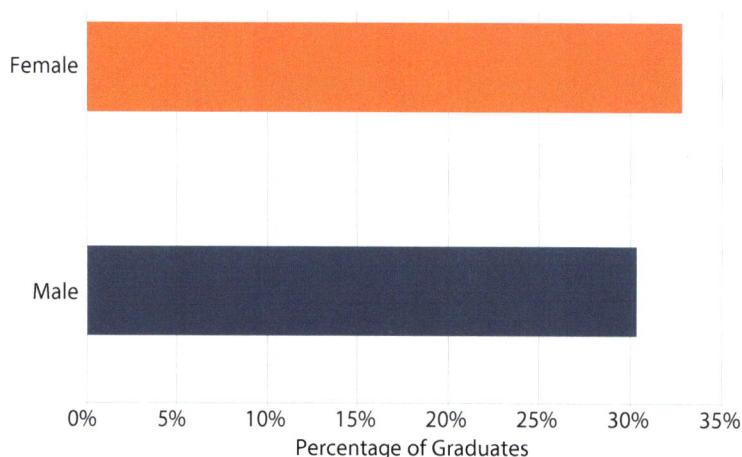

Students graduating with an undergraduate degree

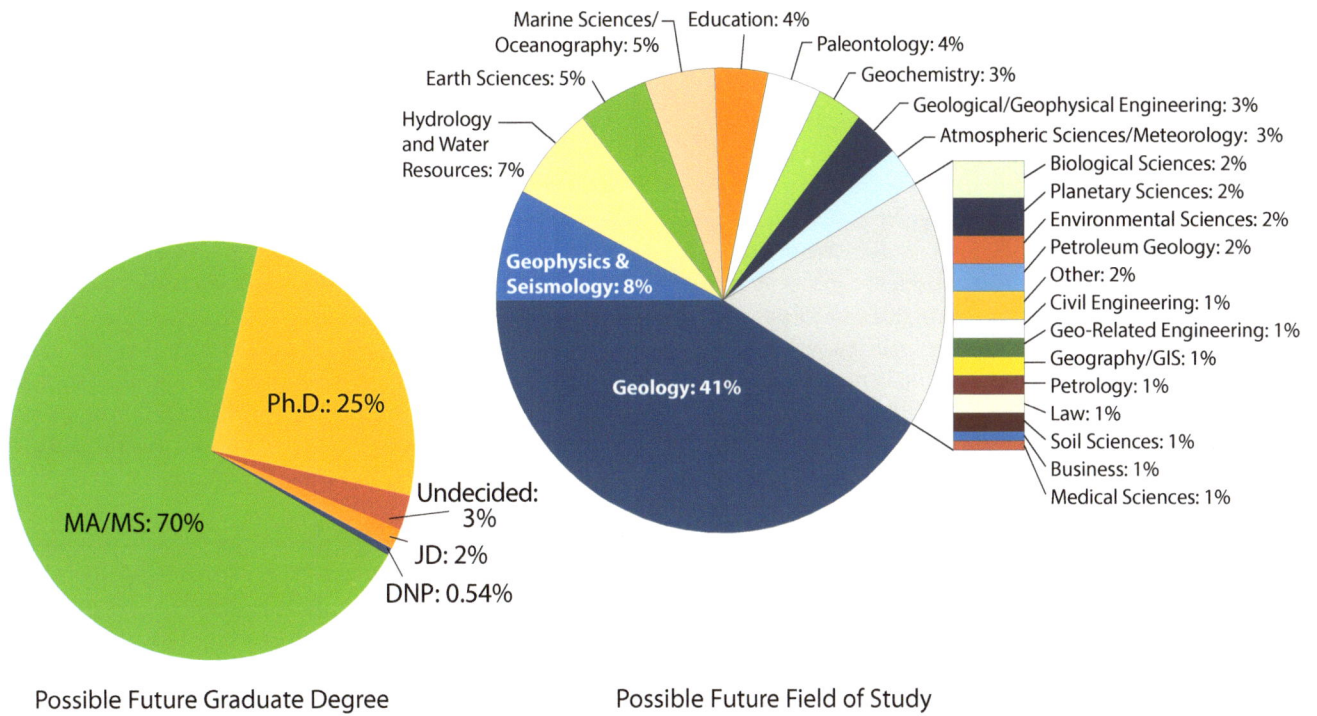

Possible Future Graduate Degree

Possible Future Field of Study

Students graduating with a graduate degree

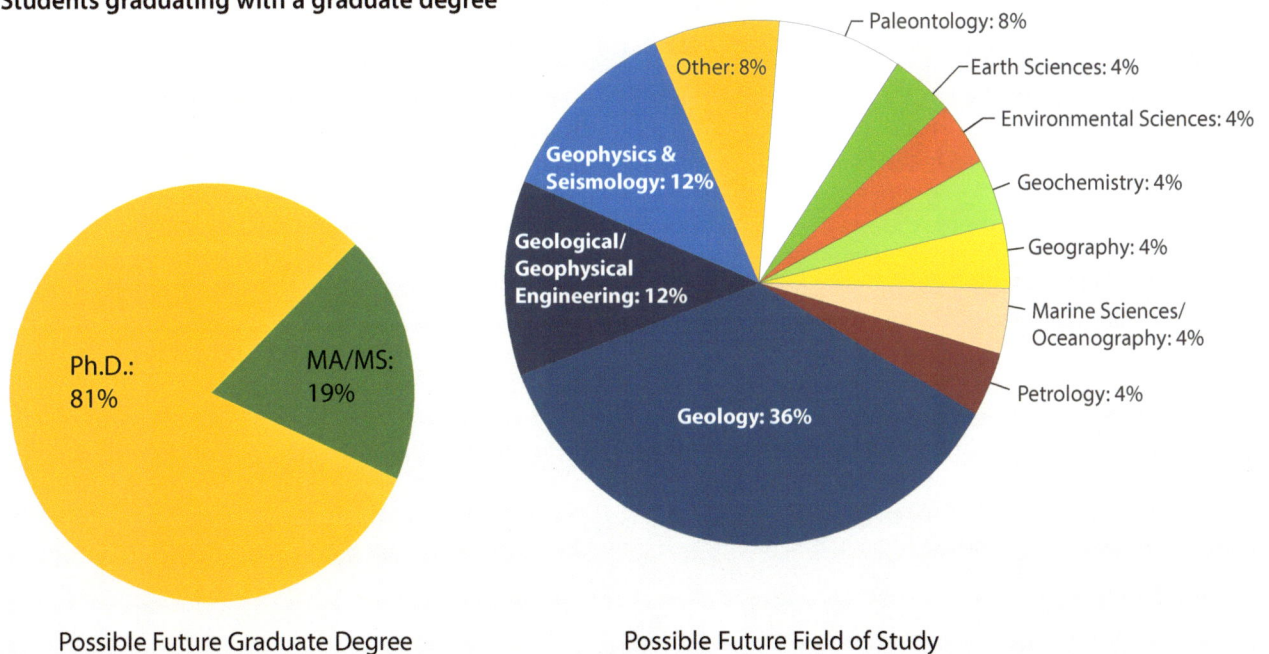

Possible Future Graduate Degree

Possible Future Field of Study

Future Plans: Working in the Geosciences

The graduates were asked if they had accepted or were seeking a job position within the geoscience workforce. If they had accepted a job, they were asked questions about these accepted job positions. Because the graduates take this survey right around graduation, it is not surprising that there are still relatively high percentages of graduates at all degree levels still seeking employment. In 2015, there was a decrease of doctoral graduates that had found a geoscience job at the time of graduation from 70 percent in 2014 to 59 percent in 2015. However, there was an increase in master's graduates that did find a geoscience job from 35 percent in 2014 to 41 percent in 2015.

This was the first year since starting AGI's Geoscience Student Exit Survey that an industry other than the oil and gas industry hired the highest percentage of bachelor's graduates. Approximately 40 percent of bachelor's graduates found a job in the environmental services industry, which was a 15 percent increase from 2014. The percentage of doctoral graduates hired by the oil and gas industry also decreased from 26 percent in 2014 to 15 percent in 2015. However, there was an increase in the percentage of master's graduates that were hired by the oil and gas industry from 59 percent in 2014 to 67 percent in 2015. The top three industries hiring bachelor's degree are environmental services, oil and gas, and the federal government. The top three industries hiring master's graduates are oil and gas, environmental services, and four-year universities. The top three industries hiring doctoral graduates are four-year universities, research institutes, and the oil and gas industry.

Over the past year, there have been changes in the employment dynamics in the oil and gas industry with an increased number of employees laid off from their positions. There have been some questions raised about a decrease in hiring of recent geoscience graduates. While there were changes in the proportion of students hired by the oil and gas industry within each degree level, it does not appear there was a drastic change to their hiring of recent graduates. However, for bachelor's graduates, it appears the jobs located within the environmental services industry are becoming a viable option to find jobs.

Most bachelor's graduates continued to find jobs with an annual salary between $20,000 and $70,000, with most of these graduates earning closer to the lower end of that range. While master's and doctoral graduates may find jobs with salaries ranging anywhere from below $30,000 to more than $120,000, the highest percentage of master's graduates fell in the salary range of $100,000–$110,000 and the highest percentage of doctoral graduates fell in the salary range of $60,000–$70,000. As in past years, every recent graduate with a starting annual salary of more than $90,000 found their job in the oil and gas industry. However, compared to 2014, there are fewer graduates making more than $110,000 per year in 2015. Starting salaries for the oil and gas industry appear to have lowered a bit as a response to the economic changes within the industry.

Graduates that found geoscience employment were asked to identify the resources they used to find their job. Bachelor's graduates tended to rely on their personal contacts, faculty referrals, and internet searches. Nearly half of the bachelor's graduates that found a geoscience job were successful at the campus recruiting events or job fairs held by the university. Doctoral graduates relied on their personal contacts mostly, but they also used the internet to locate possible jobs.

In 2015, the graduates at all degree levels still seeking employment in the geosciences showed interest in the same industries that tended to hire their peers, except for the high interest among master's graduates in state and local government positions.

The circular figure displays the connection between the degree fields of recent geoscience graduates in 2013, 2014, and 2015 (in color) to the industries where these geoscientists sound their first job after graduation (in gray). The size of the bars along the outer edge of the circle represents the number of recent graduates that pursued a particular degree field and entered a particular industry. Each colored, inner ribbon connects a particular degree field with a job in a particular industry. This visualization shows the variety of industries available to graduates with a geoscience degree, as well as the complexity of the workforce and knowledge needed in the distinct industries.

Industries where graduating students have accepted a job within the geosciences

Graduates with a BA/BS

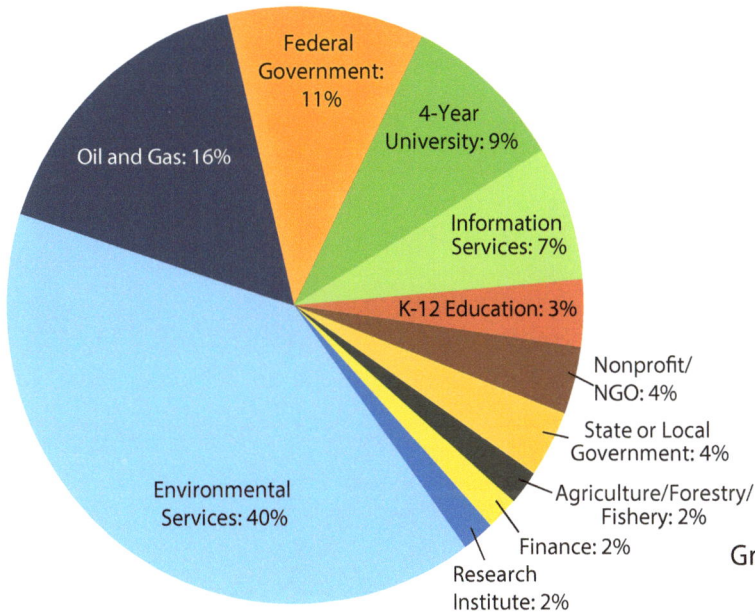

- Federal Government: 11%
- 4-Year University: 9%
- Information Services: 7%
- K-12 Education: 3%
- Oil and Gas: 16%
- Nonprofit/NGO: 4%
- State or Local Government: 4%
- Agriculture/Forestry/Fishery: 2%
- Finance: 2%
- Research Institute: 2%
- Environmental Services: 40%

Graduates with a MA/MS

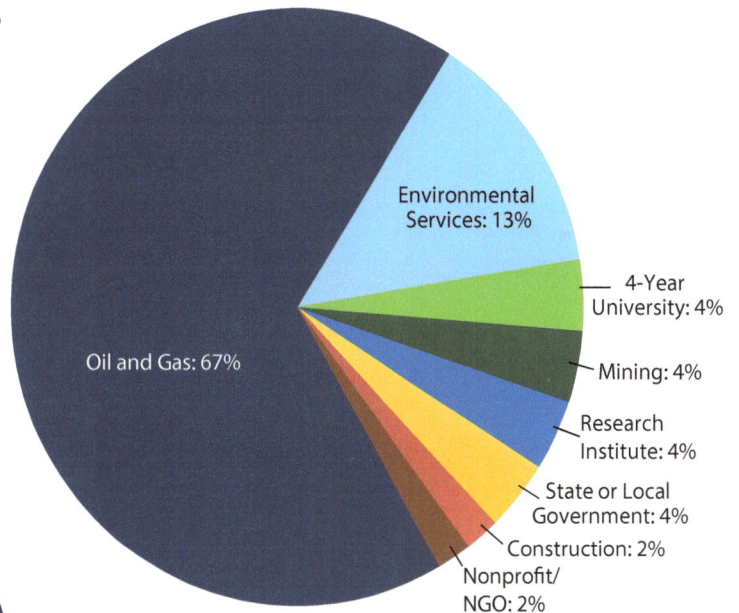

- Environmental Services: 13%
- 4-Year University: 4%
- Mining: 4%
- Research Institute: 4%
- State or Local Government: 4%
- Construction: 2%
- Nonprofit/NGO: 2%
- Oil and Gas: 67%

Graduates with a Ph.D.

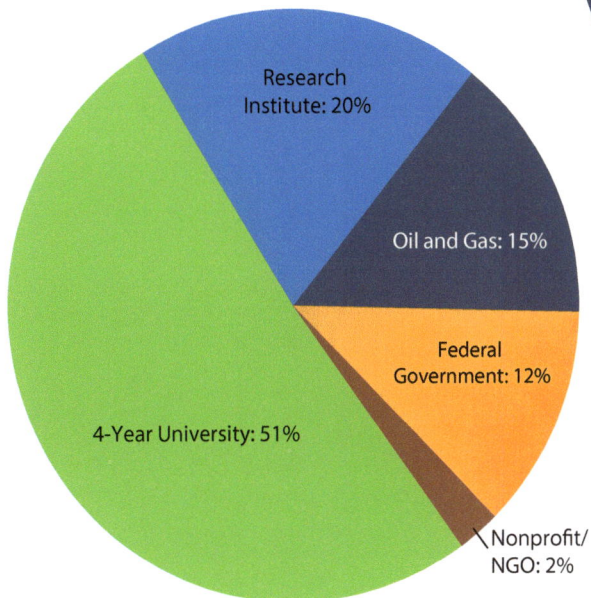

- Research Institute: 20%
- Oil and Gas: 15%
- Federal Government: 12%
- 4-Year University: 51%
- Nonprofit/NGO: 2%

Graduate students seeking or have accepted a position within the geosciences

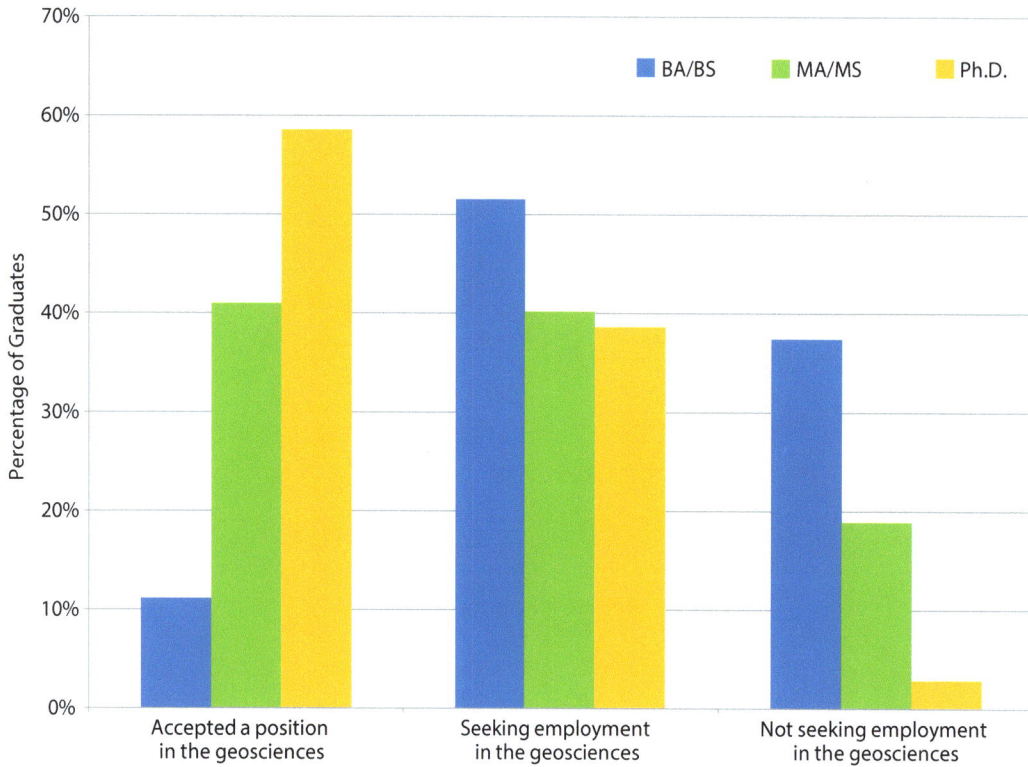

Graduate students seeking or have accepted a job within the geosciences by gender

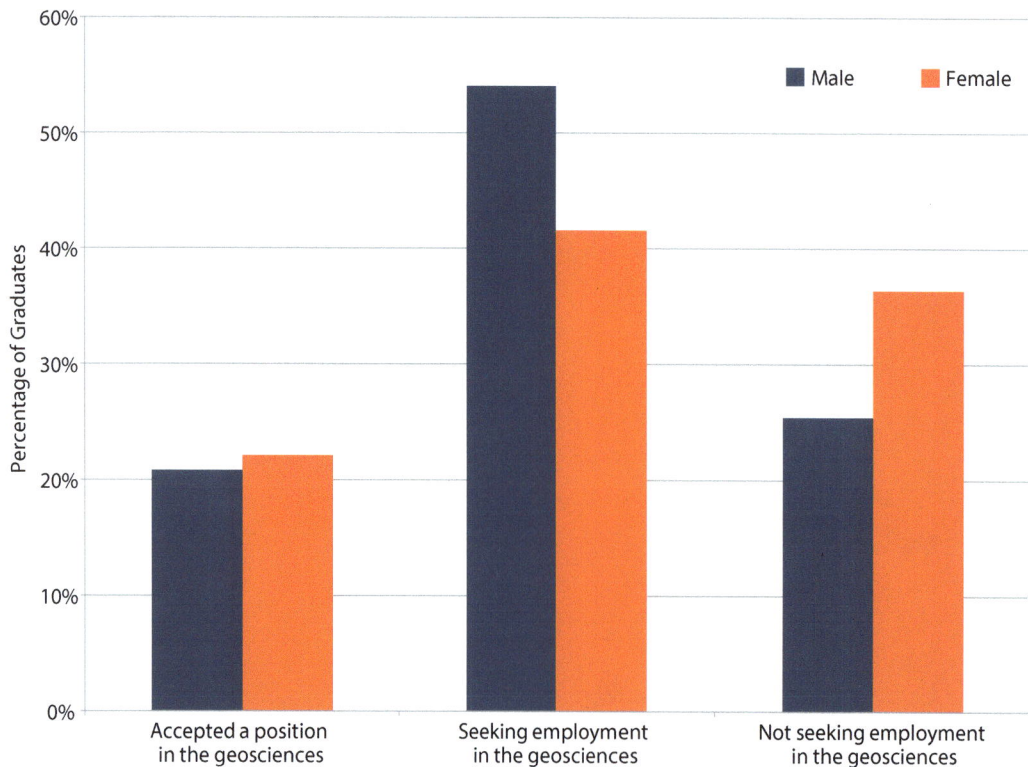

Starting salaries for graduates who accepted a job in the geosciences

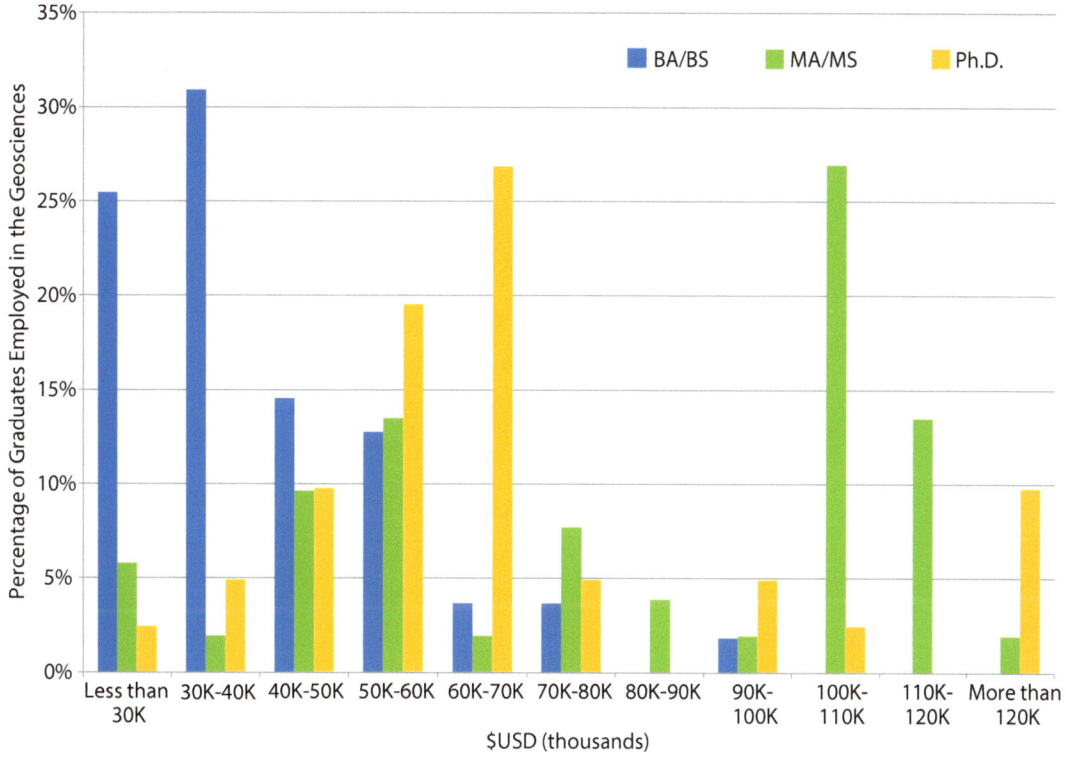

Additional compensation for graduates who accepted a job in the geosciences

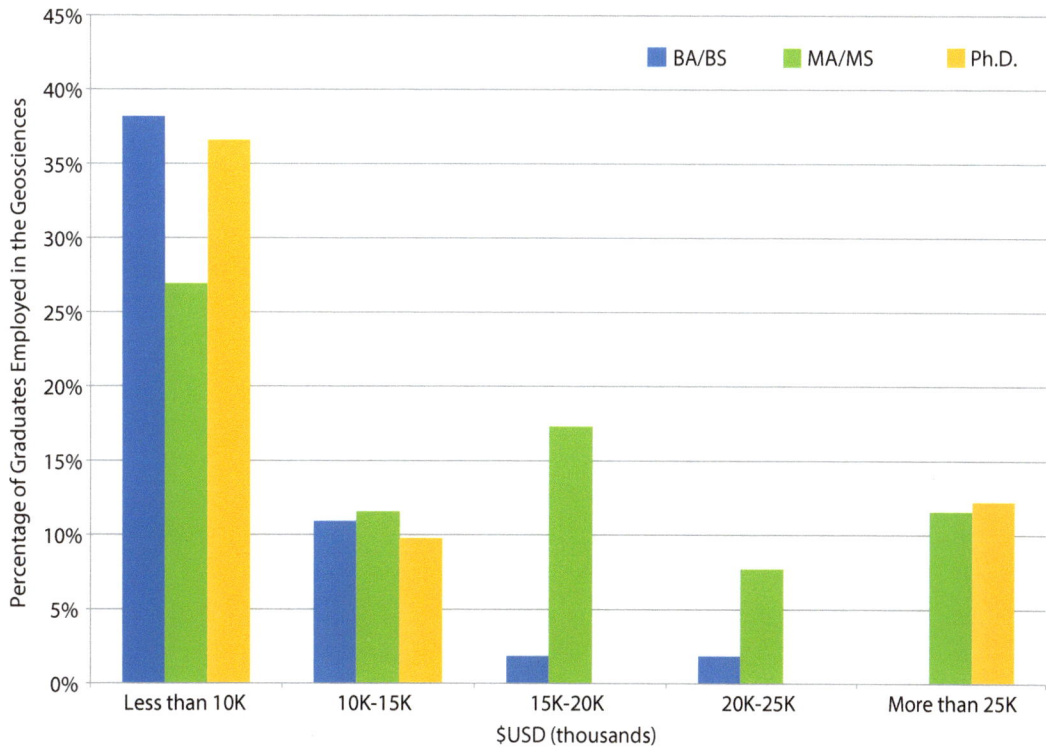

Resources identified by students as useful for finding geoscience jobs

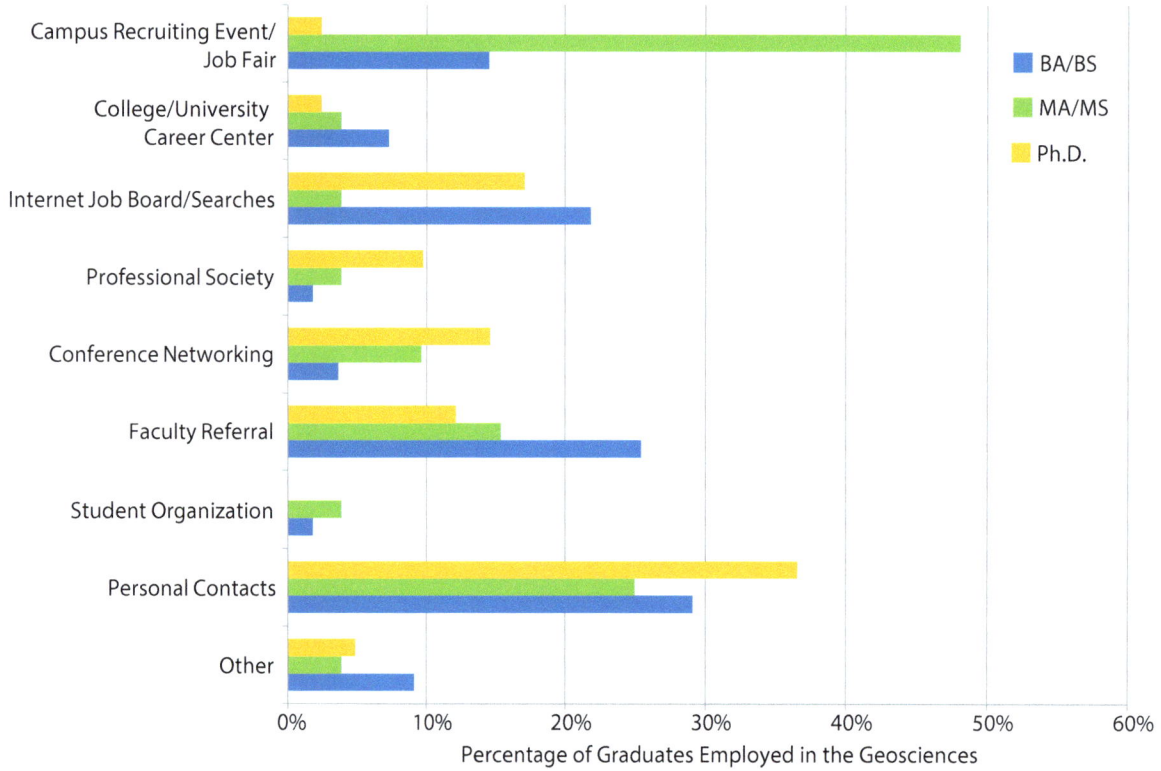

Other job opportunities offered to graduates who accepted a job in the geosciences

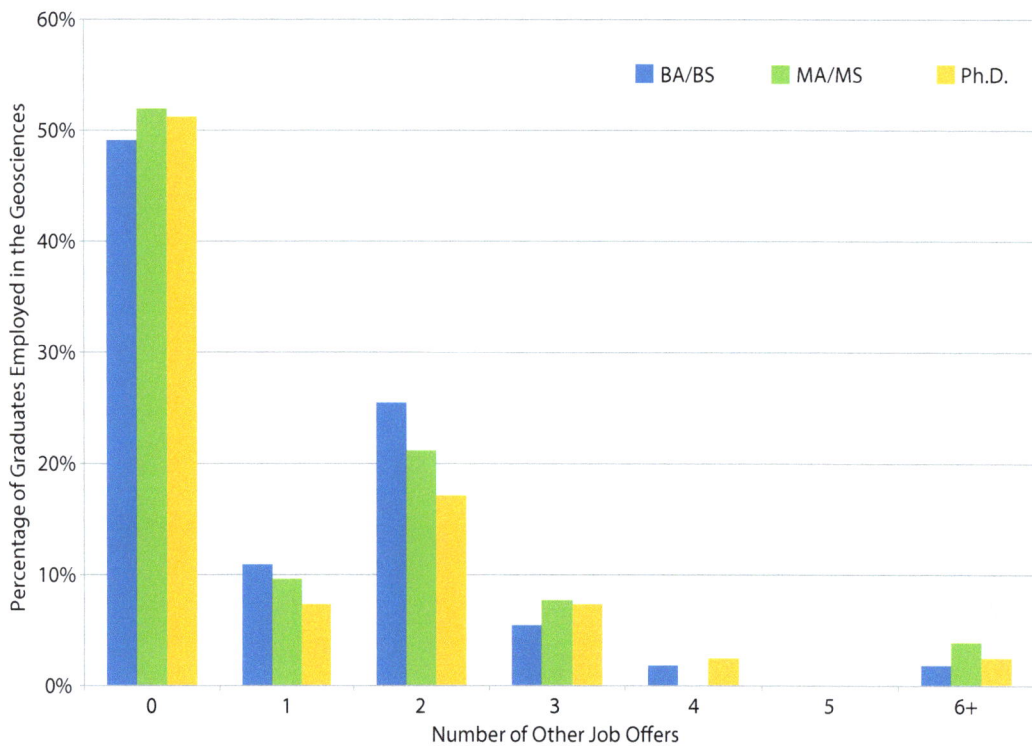

States where graduates found employment in the geosciences

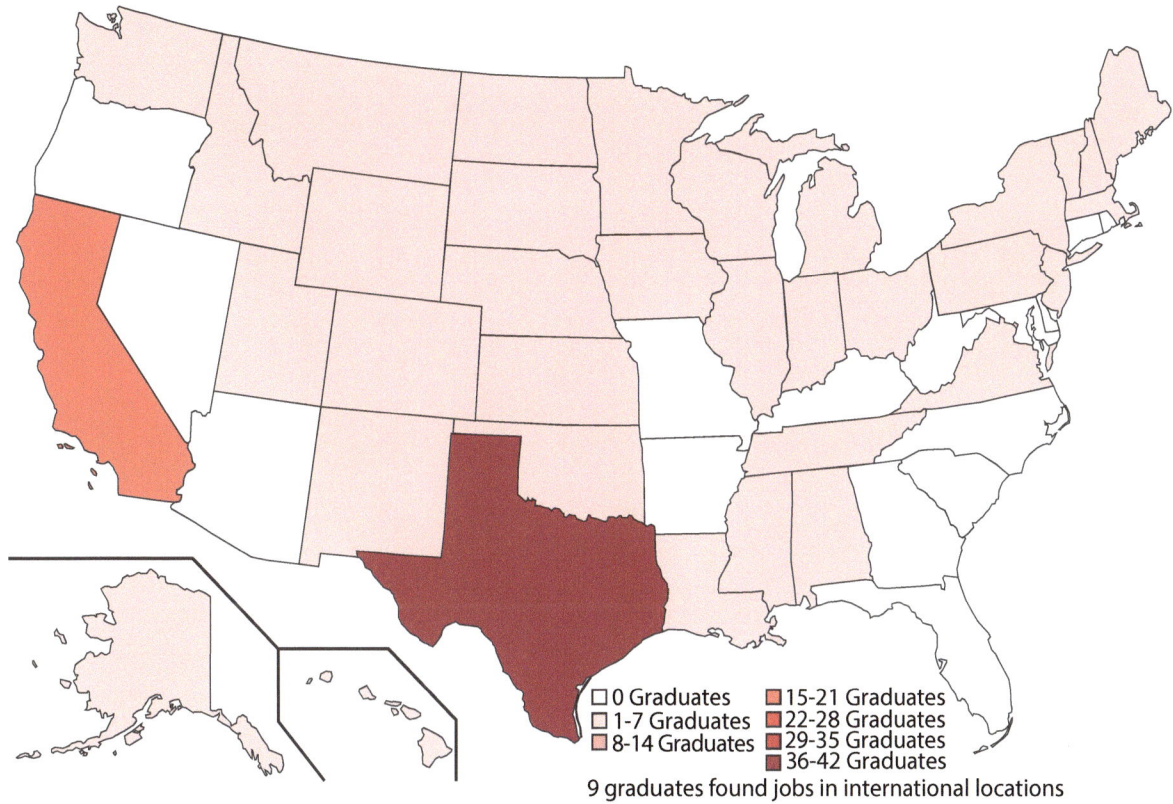

☐ 0 Graduates   🟧 15-21 Graduates
☐ 1-7 Graduates   🟥 22-28 Graduates
🟫 8-14 Graduates   🟥 29-35 Graduates
🟥 36-42 Graduates

9 graduates found jobs in international locations

Industries of interest for graduating students seeking a job within the geosciences

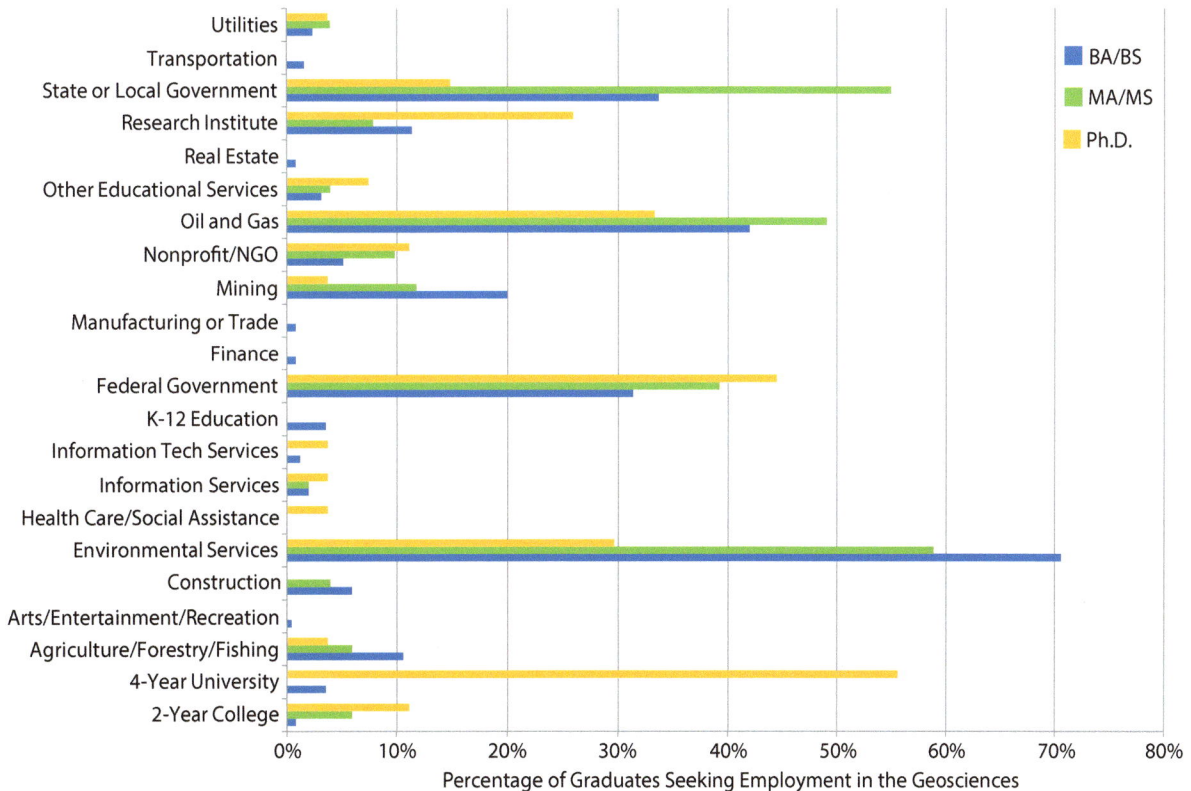

Percentage of Graduates Seeking Employment in the Geosciences

Industries of geoscience graduates' first jobs by degree field for the past three years***

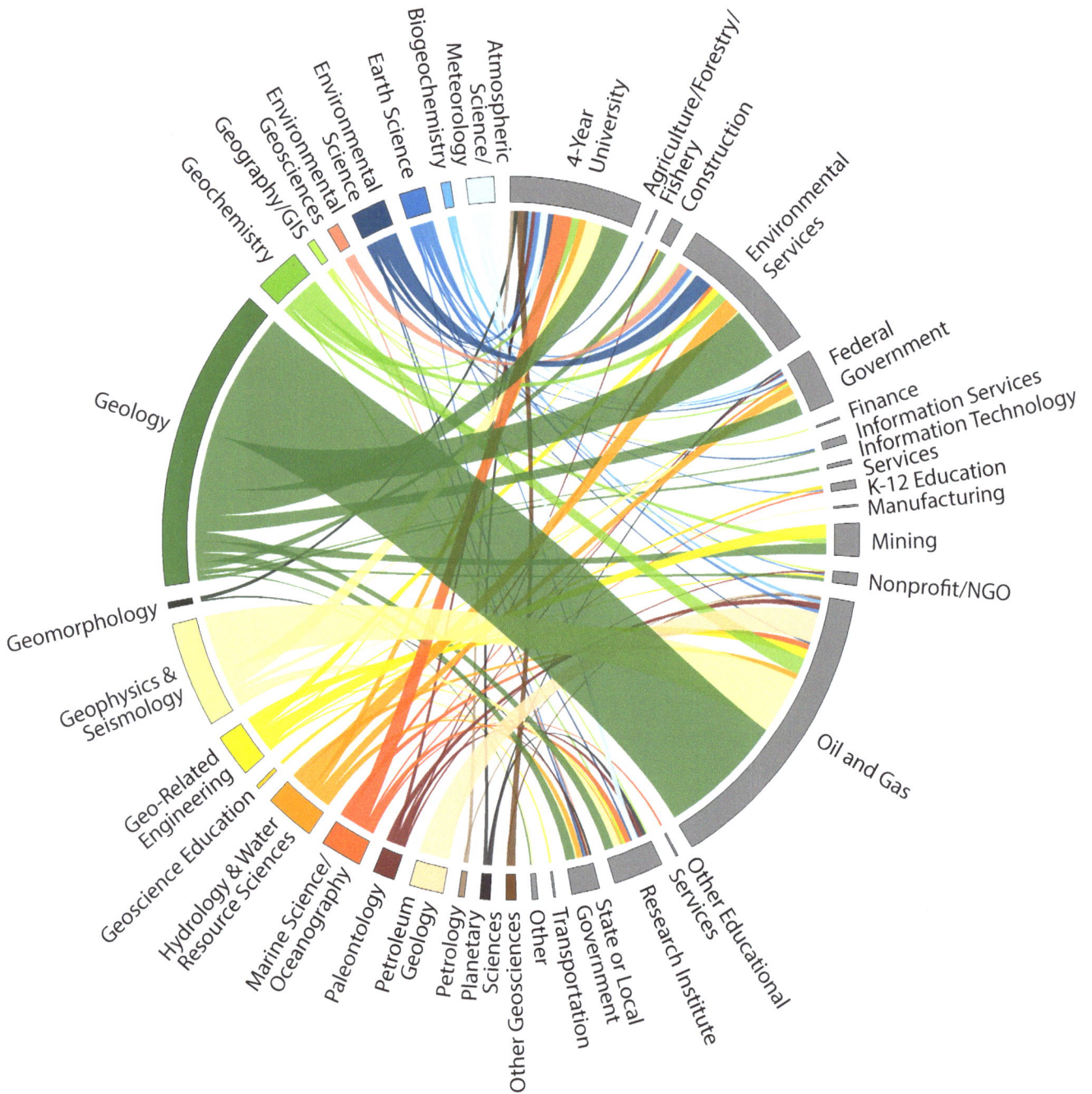

***The code for this visualization was modified from Kyzywinski, M. et al. Circos: an Information Aesthetic for Comparative Genomics. **Genome Res** (2009) 19:1693–1645

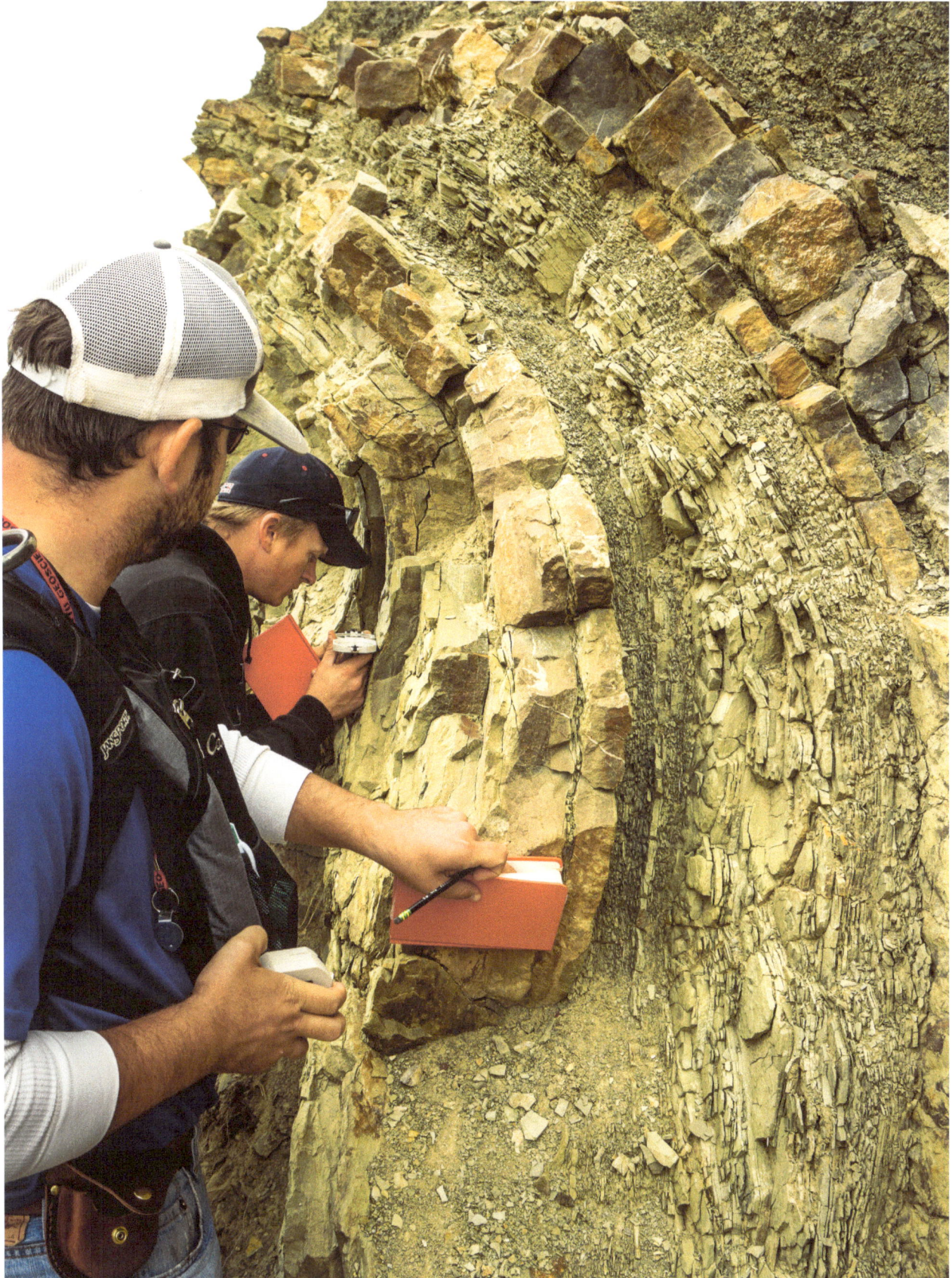

Photo by Hannah Cunningham from AGI's 2015 Life in the Field contest.
Two students measure a plunging fold.

Future Plans: Working Outside of the Geosciences

Very few students are seeking or have secured jobs outside of the geosciences. Due to this, the data about these graduates that either accepted or are seeking a job outside of the geosciences show the number of graduates regardless of degree level. Most of the graduates that have accepted a job position outside of the geosciences chose these positions because they wanted to pursue other interests, wanted a geoscience job but had trouble getting hired, and/or needed to earn money to help pay student loans or other life expenses. A few graduates noted issues with discrimination while working towards their degree or during the job application process. A couple of graduates commented on their desire for a science education job teaching earth sciences or a job in an area that allows the use of their geoscience knowledge. While these types of positions may not be considered traditional geoscience careers, AGI considers them still within the geosciences workforce.

Those graduates that had accepted a job outside of the geosciences were asked to provide more details about their jobs. The industries hiring these students into jobs outside of the geosciences were K–12 education, health care/social assistance, construction, and retail. The majority of these graduates are offered starting salaries less than $30,000 annually, and most of these jobs were found through personal contacts and internet searches.

Graduating students seeking or have accepted a job position outside the geosciences

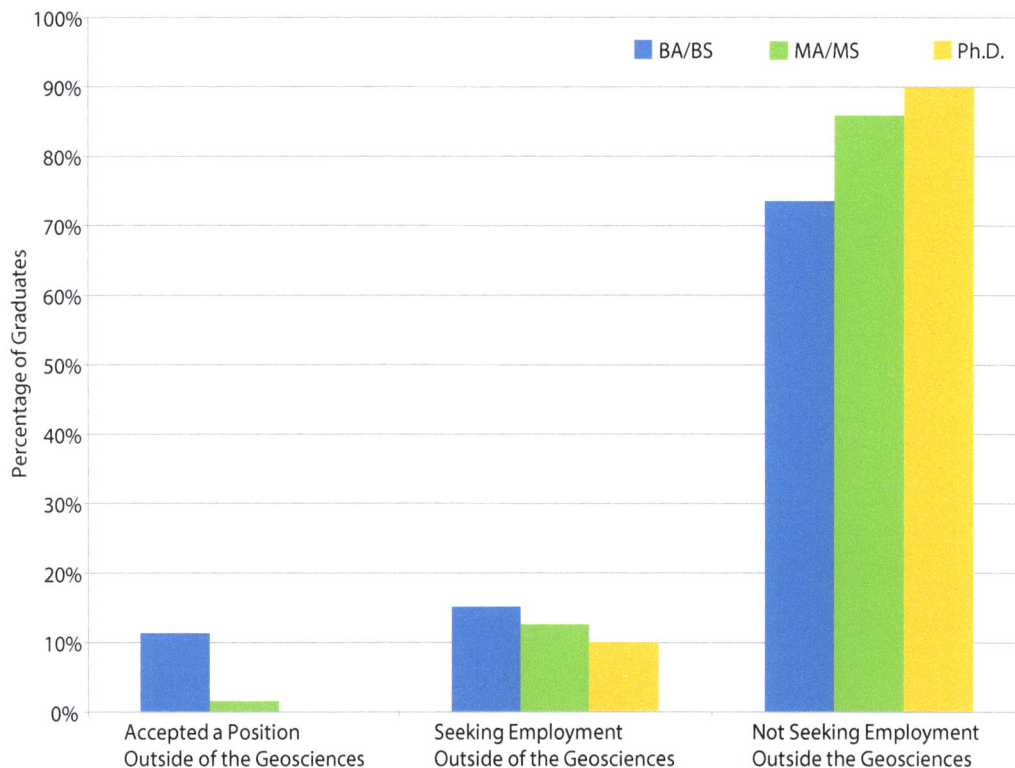

Industries where graduating students have accepted a job outside the geosciences

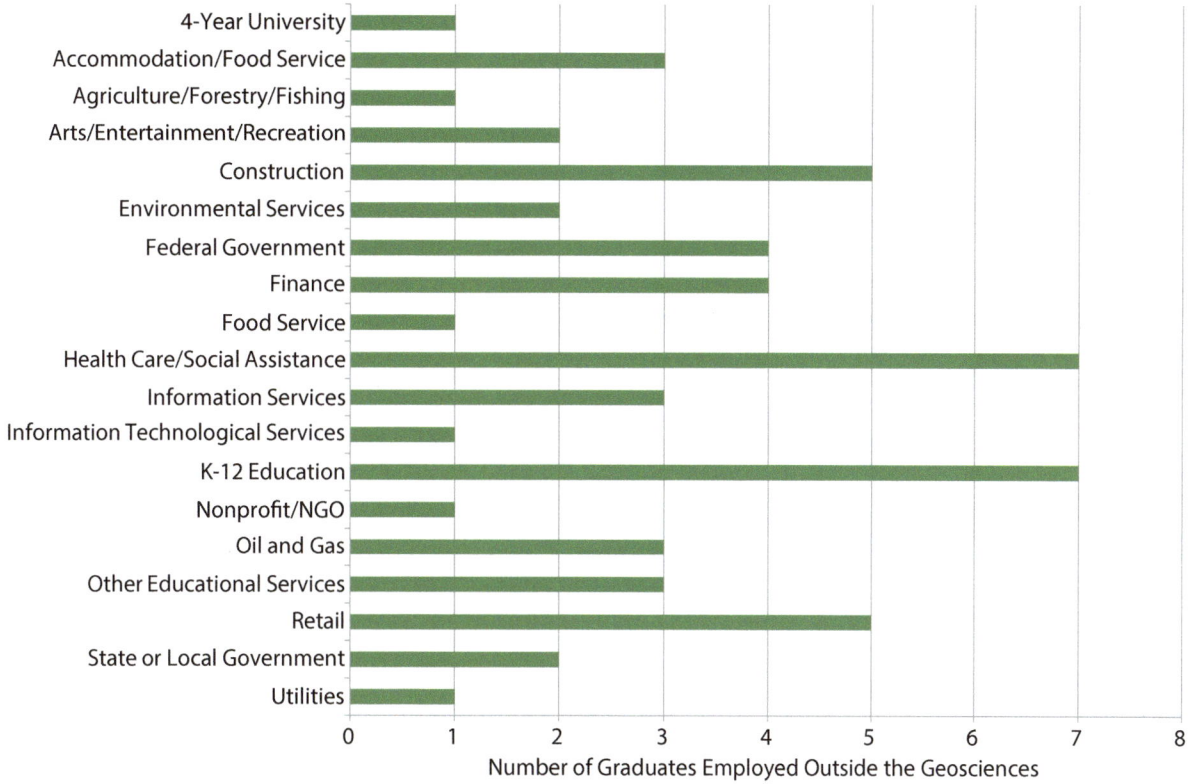

Number of Graduates Employed Outside the Geosciences

Starting salaries for graduating students that accepted a job outside the geosciences

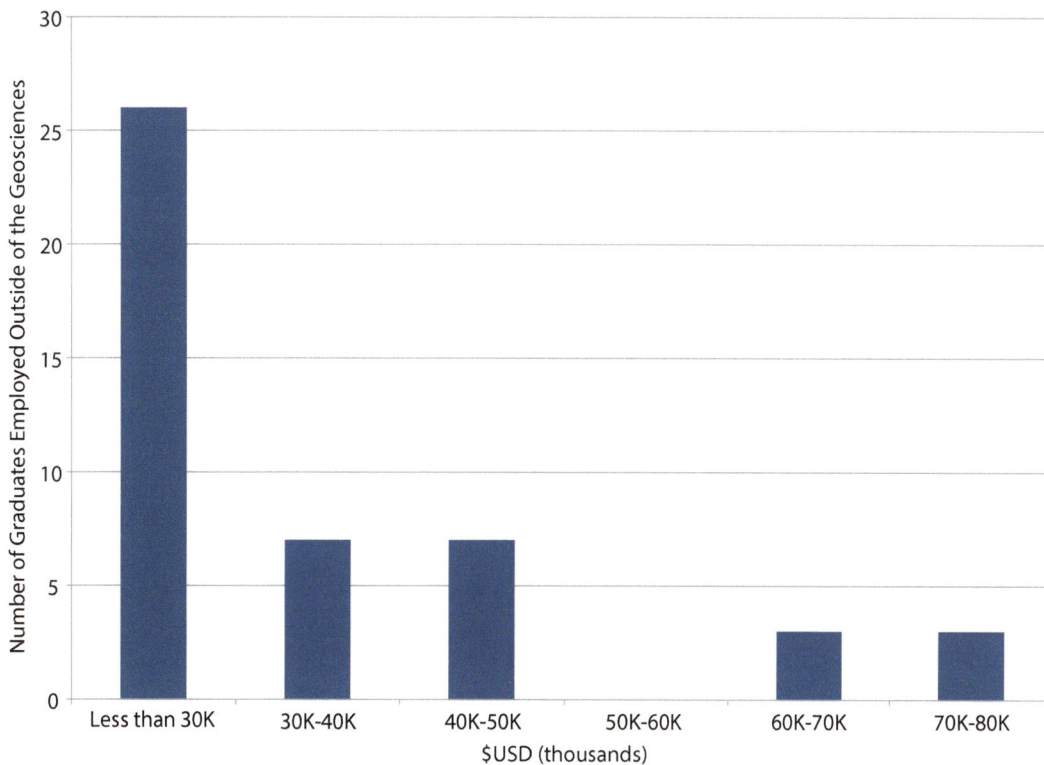

$USD (thousands)

Resources identified by graduating students as useful for finding non-geoscience jobs

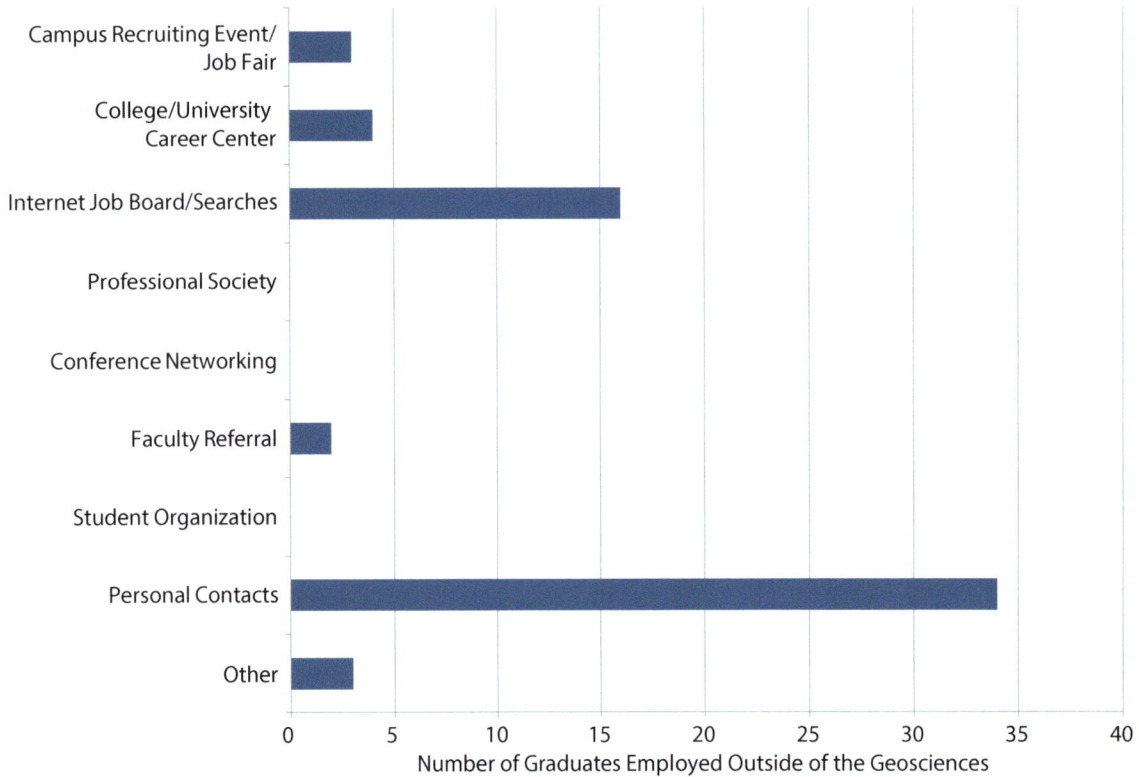

Number of Graduates Employed Outside of the Geosciences

Photo by Douglas H. Clark from AGI's 2015 Life in the Field contest.

The location is 3,000 m up on Combatant Col (next to Mt. Waddington), in the B.C. Coast Mountains. The geologist is holding on of the ice cores we collected from the col as part of an NSF-funded project to test the viability of collecting ice-core records from temperate mountain glaciers.

Photo by Tiffany Rivera from AGI's 2015 Life in the Field contest.
Students teaching in the field on a stormy day at Craters of the Moon National Monument, Idaho.

Appendices

Distribution of participating graduating students and departments*

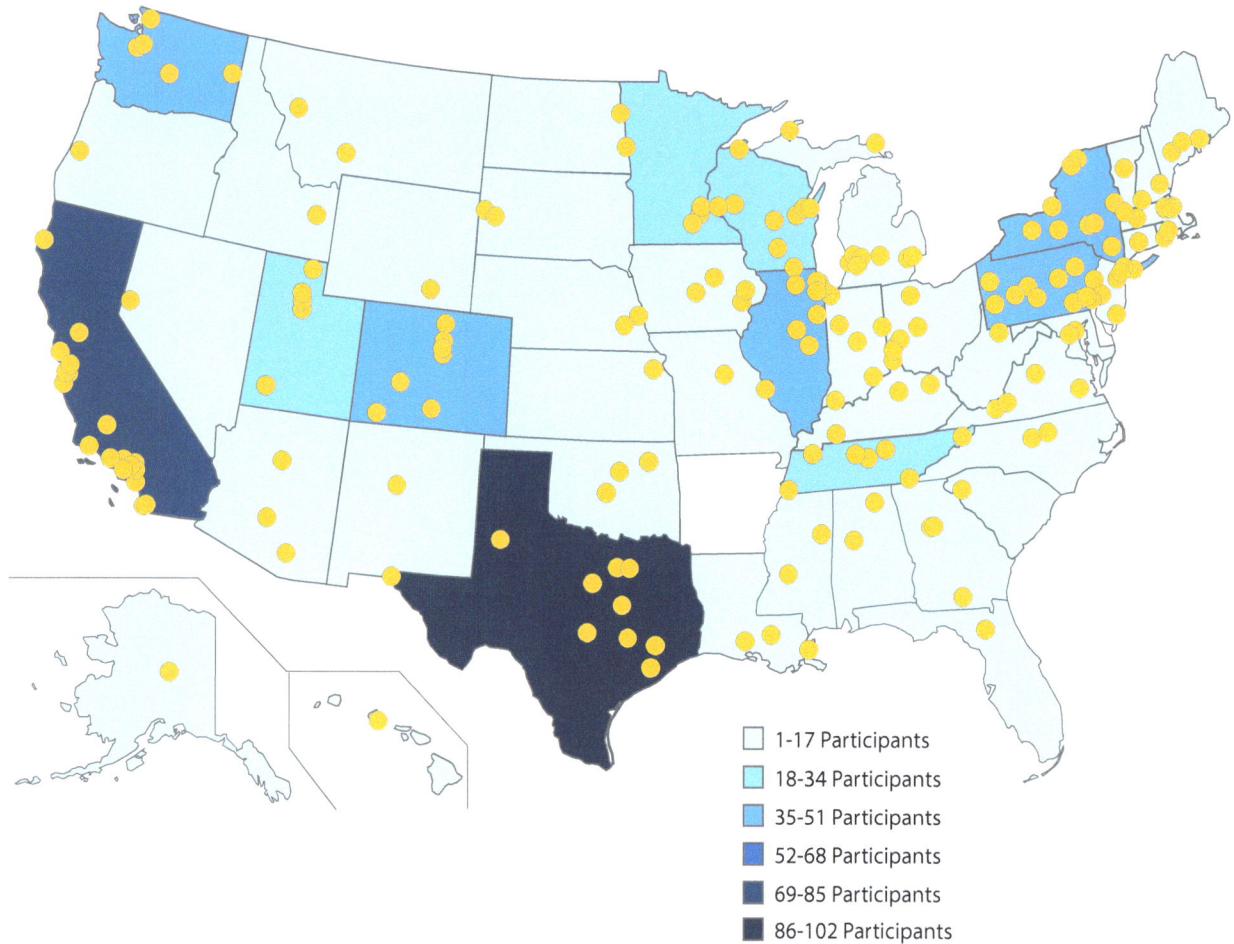

- ☐ 1-17 Participants
- ☐ 18-34 Participants
- ☐ 35-51 Participants
- ☐ 52-68 Participants
- ☐ 69-85 Participants
- ☐ 86-102 Participants

Appendix I

The following is a list of all the institutions and departments with graduating students that took AGI's Geoscience Exit Survey in the 2014–2015 academic year.

University, Department

Adams State University, Department of Biology and Earth Sciences

Amherst College, Department of Geology

Arizona State University, School of Earth and Space Exploration

Bates College, Department of Geology

Baylor University, Department of Geology

Beloit College, Department of Geology

Black Hills State University, Department of Environmental Physical Sciences

Boise State University, Department of Geosciences

Boston College, Department of Earth and Environmental Science

Boston University, Department of Earth and Environment

Bowdoin College, Department of Earth and Oceanographic Sciences

Bowling Green State University, Department of Geology

Brigham Young University, Department of Geological Sciences

Brigham Young University-Idaho, Department of Geology

Brown University, Department of Earth, Environmental and Planetary Sciences

Bryn Mawr College, Department of Geology

Bucknell University, Department of Geology and Environmental Geosciences

California Lutheran University, Department of Geology

California State Polytechnic University, Department of Geological Sciences

California State University-Bakersfield, Department of Geology

California State University-East Bay, Department of Earth and Environmental Sciences

California State University-Fullerton, Department of Geological Sciences

California State University-Long Beach, Department of Geology

California State University-Northridge, Department of Geological Sciences

Calvin College, Department of Geology, Geography, and Environmental Studies

Carleton College, Department of Geology

Central Washington University, Department of Geological Sciences

Clemson University, Department of Environmental Engineering and Earth Sciences

Colby College, Department of Geology

College of William and Mary, Department of Geology

Colorado School of Mines, Department of Civil and Environmental Engineering

Colorado School of Mines, Department of Geology and Geological Engineering

Colorado School of Mines, Department of Geophysics

Colorado State University, Department of Geosciences

Columbia University, Lamont-Doherty Earth Observatory

Cornell College, Department of Geology

Cornell University, Department of Earth and Atmospheric Sciences

Duke University, Nicholas School of the Environment

Earlham College, Department of Geology

East Tennessee State University, Department of Geosciences

Eastern Washington University, Department of Geology

Fort Lewis College, Department of Geosciences

Franklin and Marshall College, Department of Earth and Environment

Georgia Institute of Technology, School of Biology

Georgia Institute of Technology, School of Earth and Atmospheric Sciences

Georgia State University, Department of Geosciences

Grand Valley State University, Department of Geology

Gustavus Adolphus College, Department of Geology

Hanover College, Department of Geology

Hartwick College, Department of Geological and Environmental Sciences

Hofstra University, Department of Geology

Hope College, Department of Geological and Environmental Sciences

Humboldt State University, Department of Geology

Illinois State University, Department of Geology

Indiana University Northwest, Department of Geosciences

Indiana University of Pennsylvania, Department of Geoscience

Indiana University-Purdue University Fort Wayne, Department of Geosciences

Iowa State University, Department of Geological and Atmospheric Sciences

Juniata College, Department of Geology

Keene State College, Department of Geology

Lafayette College, Department of Geology and Environmental Geosciences

Lake Superior State University, Department of Geology

Lawrence University, Department of Geology

Louisiana State University, Department of Geology and Geophysics

Miami University of Ohio, Department of Geology and Environmental Earth Science

Michigan State University, Department of Geological Sciences

Michigan Technological University, Department of Geological/Mining Engineering and Sciences

Middle Tennessee State University, Department of Geosciences

Millersville University, Department of Earth Science

Millsaps College, Department of Geology

Mississippi State University, Department of Geosciences

MIT, Department of Earth, Atmospheric, and Planetary Sciences

Montana State University, Department of Geology

Montclair State University, Department of Earth and Environmental Studies

Morehead State University, Department of Geology and Earth Systems Science

North Dakota State University, Department of Geosciences

Northeastern Illinois University, Department of Earth Sciences

Northern Arizona University, School of Earth Sciences and Environmental Sustainability

Northern Illinois University, Department of Geology and Environmental Geosciences

Northland College, Department of Geosciences

Northwestern University, Department of Earth and Planetary Sciences

Norwich University, Department of Geology

Ohio State University, Department of Geography

Ohio State University, School of Earth Sciences

Oklahoma State University, Department of Geology

Olivet Nazarene University, Department of Geological Sciences

Oregon State University, College of Earth, Ocean, and Atmospheric Sciences

Pacific Lutheran University, Department of Geoscience

Pennsylvania State University, Department of Geosciences

Pomona College, Department of Geology

Purdue University, Department of Earth and Atmospheric Sciences

Radford University, Department of Geology

Rutgers University, Department of Earth and Planetary Sciences

Sam Houston State University, Department of Geography and Geology

San Diego State University, Department of Geological Sciences

Skidmore College, Department of Geosciences

Slippery Rock University, Department of Geography, Geology, and the Environment

Sonoma State University, Department of Geology

South Dakota School of Mines and Technology, Department of Geology and Geological Engineering

Southern Utah University, Department of Geology

St. Lawrence University, Department of Geology

St. Norbert College, Department of Geology

Stanford University, Department of Earth System Science

Stanford University, Department of Geophysics

Stockton University, Department of Geology

SUNY Geneseo, Department of Geological Sciences

SUNY New Paltz, Department of Geological Sciences

SUNY Potsdam, Department of Geology

SUNY Oneonta, Department of Earth and Atmospheric Sciences

SUNY Oswego, Department of Atmospheric and Geological Sciences

Tarleton State University, Department of Geoscience

Temple University, Department of Earth and Environmental Science

Tennessee Tech University, Department of Earth Sciences

Texas A&M University, School of Geosciences

Texas Tech University, Department of Geosciences

Tulane University, Department of Earth and Environmental Sciences

University of Alabama, Department of Geological Sciences

University of Alaska-Fairbanks, Department of Geosciences

University of Alaska-Fairbanks, Department of Atmospheric Sciences

University of Alaska-Fairbanks, College of Engineering and Mines

University of Arizona, Department of Geosciences

University of California-Berkeley, Department of Earth and Planetary Science

University of California-Davis, Department of Earth and Planetary Sciences

University of California-Los Angeles, Department of Earth, Planetary and Space Sciences

University of California-San Diego, Scripps Institution of Oceanography

University of California-Santa Barbara, Department of Earth Science

University of California-Santa Cruz, Department of Earth and Planetary Sciences

University of Cincinnati, Department of Geology

University of Colorado at Boulder, Department of Geological Sciences

University of Colorado at Boulder, Environmental Studies Program

University of Denver, Department of Geography and the Environment

University of Florida, Department of Geological Sciences

University of Florida, School of Natural Resources and Environment

University of Hawaii-Manoa, Department of Biology

University of Hawaii-Manoa, School of Ocean & Earth Science & Technology

University of Houston, Department of Earth and Atmospheric Sciences

University of Illinois at Chicago, Department of Earth and Environmental Sciences

University of Illinois, Department of Geology

University of Iowa, Department of Earth and Environmental Sciences

University of Kansas, Department of Geology

University of Kentucky, Department of Earth and Environmental Sciences

University of Louisiana at Lafayette, Department of Geology

University of Maryland, Department of Geology

University of Maryland-Baltimore County, Department of Geography and Environmental Systems

University of Massachusetts, Department of Geosciences

University of Memphis, Department of Earth Sciences

University of Michigan, Department of Earth and Environmental Sciences

University of Minnesota, Department of Civil, Environmental, and Geo Engineering

University of Minnesota, Department of Soil, Water and Climate

University of Minnesota, Department of Earth Sciences

University of Minnesota, Department of Forest Resources

University of Missouri, Department of Geological Sciences

University of Montana, Department of Geosciences

University of Nebraska-Lincoln, Department of Earth and Atmospheric Sciences

University of Nebraska-Omaha, Department of Geography/ Geology

University of Nevada-Reno, Mackay School of Earth Science and Engineering

University of New Hampshire, Institute for the Study of Earth, Oceans, and Space

University of New Mexico, Department of Earth and Planetary Sciences

University of North Carolina at Chapel Hill, Department of Geosciences

University of North Carolina at Wilmington, Department of Geography and Geology

University of North Dakota, School of Geology and Geological Engineering

University of North Georgia, Institute for Environmental and Spatial Analysis

University of Northern Iowa, Department of Earth Science

University of Oklahoma, School of Geology and Geophysics

University of Pennsylvania, Department of Earth and Environmental Science

University of Pittsburgh, Department of Geology and Planetary Sciences

University of Rhode Island, Department of Geosciences

University of Southern Indiana, Department of Geology

University of St. Thomas, Department of Geology

University of Tennessee at Chattanooga, Department of Geology

University of Tennessee at Martin, Department of Agriculture and Applied Sciences

University of Texas at Arlington, Department of Earth and Environmental Sciences

University of Texas at Austin, Jackson School of Geosciences

University of Texas at Dallas, Department of Geosciences

University of Texas at El Paso, Department of Geological Sciences

University of Tulsa, Department of Geosciences

University of Utah, College of Mines and Earth Sciences

University of Virginia, Department of Environmental Sciences

University of Washington, Department of Earth and Space Sciences

University of Washington, Department of Oceanography

University of Wisconsin-Green Bay, Department of Natural and Applied Sciences

University of Wisconsin-Green Bay, Environmental Science & Policy Program

University of Wisconsin-Eau Claire, Department of Geology

University of Wisconsin-Madison, Department of Geology and Geophysics

University of Wisconsin-River Falls, Department of Geology

University of Wisconsin-Stevens Point, Department of Geography and Geology

University of Wyoming, Department of Geology and Geophysics

Utah State University, Department of Geology

Valdosta State University, Department of Physics, Astronomy, and Geosciences

Vanderbilt University, Department of Earth and Environmental Sciences

Virginia Polytechnic Institute and State University, Department of Geosciences

Washington University in St. Louis, Department of Earth and Planetary Sciences

Wayne State University, Department of Geology

Weber State University, Department of Geosciences

Wesleyan University, Department of Earth and Environmental Sciences

West Chester University, Department of Geology and Astronomy

West Virginia University, Department of Geology and Geography

Western Kentucky University, Department of Geography and Geology

Western Michigan University, Department of Geosciences

Western State Colorado University, Department of Geology

Western Washington University, Department of Geology

Wheaton College, Department of Geology and Environmental Science

Wilkes University, Department of Environmental Engineering and Earth Sciences

Williams College, Department of Geosciences

Wittenberg University, Department of Geology

Wright State University, Department of Earth and Environmental Sciences

Appendix II

Carnegie Classifications of Institutions of Higher Learning
(http://carnegieclassifications.iu.edu//resources/links.php)

This classification system was used for some of the analysis of the spring 2015 results of AGI's Geoscience Student Exit Survey. The following are the definitions for the classification system and the participating institutions belonging to each category as defined and categorized by the Carnegie Foundation for the Advancement of Teaching.

Baccalaureate Colleges — Arts & Sciences (Bac/A&S)

Baccalaureate Colleges — Diverse Fields (Bac/Diverse)

Includes institutions where baccalaureate degrees represent at least 10 percent of all undergraduate degrees and where fewer than 50 master's degrees or 20 doctoral degrees were awarded during the update year. Excludes Special Focus Institutions and Tribal Colleges.

Among Institutions where bachelor's degrees represented at least half of all undergraduate degrees, those with at least half of bachelor's degree majors in arts and science fields were included in the "Arts & Sciences" group, while the remaining institutions were included in the "Diverse Fields" group.

Exit Survey Departments (Bac/A&S):
Amherst College
Bates College
Beloit college
Bowdoin College
Bryn Mawr College
Bucknell University
Calvin College
Carleton College
Colby College
Cornell College
Earlham College
Fort Lewis College
Franklin and Marshall College
Gustavus Adolphus College
Hanover College
Hartwick College

Hope College
Juniata College
Lafayette College
Lawrence University
Millsaps College
Northland College
Pomona College
Skidmore College
St. Lawrence University
St. Norbert College
Wesleyan University
Western State Colorado University
Wheaton College
Williams College
Wittenberg University

Exit Survey Departments (Bac/Diverse)*
Brigham Young University-Idaho
Lake Superior State University

Master's Colleges and Universities — Larger Programs (Master's/L)

Master's Colleges and Universities — Medium Programs (Master's/M)

Master's Colleges and Universities — Smaller Programs (Master's/S)

Generally includes institutions that awarded at least 50 master's degrees and fewer than 20 doctoral degrees during the update year (with occasional exceptions). Excludes Special Focus Institutions and Tribal Colleges.

Master's program size was based on the number of master's degrees awarded during the update year. Those awarding at least 200 degrees were included among larger programs; those awarding 100-199 were included among the medium programs; and those awarding 50-99 were included among the smaller programs. The smaller programs group also includes institutions that awarded fewer than 50 master's degrees if (a) their Enrollment Profile classification is Exclusively Graduate/Professional or (b) their Enrollment Profile classification is Majority Graduate/Professional and they awarded more graduate/professional degrees than undergraduate degrees.

Exit Survey Departments (Master's/L):
California Lutheran University

California State Polytechnic University
California State University-Bakersfield
California State University-East Bay
California State University-Fullerton
California State University-Long Beach
California State University-Northridge
Eastern Washington University
Grand Valley State University
Indiana University-Purdue University
Millersville University
Montclair State University
Morehead State University
Northeastern Illinois University
Norwich University
Olivet Nazarene University
Radford University
Slippery Rock University
Sonoma State University
Southern Utah University
SUNY New Paltz
SUNY Oswego
SUNY Potsdam
Tarleton State University
Tennessee Tech University
University of North Carolina at Wilmington
University of North Georgia
University of Northern Iowa
University of Southern Indiana
University of Tennessee at Chattanooga
Valdosta State University
West Chester University
Western Kentucky University
Western Washington University
Wilkes University

Exit Survey Departments (Master's/M):
Adams State University
Central Washington University
Humboldt State University
Indiana University Northwest
Pacific Lutheran University
Stockton University
University of Tennessee at Martin
University of Wisconsin-Eau Claire
University of Wisconsin-River Falls
University of Wisconsin-Stevens Point
Weber State University

Exit Survey Departments (Master's/S):
Black Hills State University
Keene State College
SUNY Geneseo
SUNY Oneonta
University of Wisconsin-Green Bay

Research Universities—
Very High Research Activity (RU/VH)

Research Universities—
High Research Activity (RU/H)

Doctoral/Research Universities (DRU)

Includes institutions that awarded at least 20 research doctoral degrees during the update year (excluding doctoral-level degrees that qualify recipients for entry into professional practice, such as the JD, MD, PharmD, DPT, etc.). Excludes Special Focus Institutions and Tribal Colleges.

Doctorate-granting institutions were assigned to one of three categories based on a measure of research activity. For more information about the analysis of research activity, please visit http://carnegieclassifications.iu.edu/methodology/basic.php.

Exit Survey Departments (RU/VH):
Arizona State University
Boston University
Brown University
Colorado State University
Columbia University
Cornell University
Duke University
Georgia Institute of Technology
Georgia State University
Iowa State University
Louisiana State University
Michigan State University
Mississippi State University
MIT
Montana State University
North Dakota State University
Northwestern University
Ohio State University
Oregon State University
Pennsylvania State University
Purdue University
Rutgers University
Stanford University
Texas A&M University
Tulane University
University of Arizona
University of California-Berkeley
University of California-Davis
University of California-Los Angeles
University of California-San Diego
University of California-Santa Barbara
University of California-Santa Cruz
University of Cincinnati
University of Colorado at Boulder
University of Florida

University of Hawaii-Manoa
University of Houston
University of Illinois at Chicago
University of Illinois
University of Iowa
University of Kansas
University of Kentucky
University of Maryland
University of Massachusetts
University of Michigan
University of Minnesota
University of Nebraska-Lincoln
University of New Mexico
University of North Carolina at Chapel Hill
University of Oklahoma
University of Pennsylvania
University of Pittsburgh
University of Texas at Austin
University of Utah
University of Virginia
University of Washington
University of Wisconsin-Madison
Vanderbilt University
Virginia Polytechnic Institute and State University
Washington University in St. Louis
Wayne State University

Exit Survey Departments (RU/H):

Baylor University
Boston College
Bowling Green State University
Brigham Young University
Clemson University
College of William and Mary
Colorado School of Mines
Miami University of Ohio
Michigan Technological University
Northern Arizona University
Northern Illinois University
Oklahoma State University
Portland State University
San Diego State University
Temple University
Texas Tech University
University of Alabama
University of Alaska-Fairbanks
University of Denver
University of Louisiana at Lafayette
University of Maryland-Baltimore County
University of Memphis
University of Missouri
University of Montana
University of Nevada-Reno
University of New Hampshire
University of North Dakota
University of Rhode Island
University of Texas at Arlington
University of Texas at Dallas

University of Texas at El Paso
University of Wyoming
Utah State University
West Virginia University
Western Michigan University
Wright State University

Exit Survey Departments (DRU):

East Tennessee State University
Hofstra University
Illinois State University
Indiana University of Pennsylvania
Middle Tennessee State University
Sam Houston State University
University of Nebraska-Omaha
University of St. Thomas
University of Tulsa

Special Focus Institutions— Schools of Engineering (Spec/Engg)

Institutions awarding baccalaureate or higher-level degrees where a high concentration of degrees (above 75%) is in a single field or set of related fields. Excludes Tribal Colleges.

Exit Survey Departments (Spec/Engg)*:

South Dakota School of Mines and Technology

*Institutions in this classification where not included in comparisons using the Carnegie Classification system due to the small number of institutions in the Exit Survey belonging to the particular classification.

www.ingramcontent.com/pod-product-compliance
Lightning Source LLC
Chambersburg PA
CBHW052055190326
41519CB00002BA/228